T0186312

EUROCK '96
VOLUME/TOME/BAND 3

EUROCK '96/TORINO/ITALY
PROCEEDINGS/COMPTES-RENDUS/SITZUNGSBERICHTE
ISRM INTERNATIONAL SYMPOSIUM/1996.09.2-5

Prediction and Performance
in Rock Mechanics and Rock Engineering

Prévisions et réalisations
en Mécanique et Ingéniérie des Roches

Voraussichten und Leistungen
in Felsmechanik und Felsingenieurwesen

Editor/Editeur/Herausgeber
GIOVANNI BARLA
Department of Structural Engineering, Politecnico di Torino, Italy

VOLUME/TOME/BAND 3

Published for/Publiés pour/Herausgegeben für
AGI – Associazione Geotecnica Italiana

A.A.BALKEMA/ROTTERDAM/BROOKFIELD/2000

The Associazione Geotecnica Italiana and the Department of Structural Engineering of the Politecnico di Torino

are especially grateful to

CITTÁ DI TORINO
PROVINCIA DI TORINO
REGIONE PIEMONTE
POLITECNICO DI TORINO

AEM, TORINO
AGIP S.P.A., MILANO
CASSA DI RISPARMIO DI TORINO
S.E.L.I. S.P.A., ROMA

Cover: Large storage cavern in jointed chalk. Comparison of predicted and measured displacements during staged excavation.
Courtesy of G. Barla

Authorization to photocopy items for internal or personal use, or the internal or personal use of specific clients, is granted by A.A. Balkema, Rotterdam, provided that the base fee of US$1.50 per copy, plus US$0.10 per page is paid directly to Copyright Clearance Center, 222 Rosewood Drive, Danvers, MA 01923. For those organizations that have been granted a photocopy license by CCC, a separate system of payment has been arranged. The fee code for users of the Transactional Reporting Service is: 90 5410 843 6/00 US$1.50 + US$0.10.

The texts of the various papers in this volume were set individually by typists under the supervision of each of the authors concerned.

Complete set of three volumes: ISBN 90 5410 843 6
Volume 1: ISBN 90 5410 841 X
Volume 2: ISBN 90 5410 842 8
Volume 3: ISBN 90 5410 844 4

Published by:
© 2000 A.A. Balkema, P.O. Box 1675, 3000 BR Rotterdam, Netherlands
Fax: +31.10.413.5947; balkema@balkema.nl; www.balkema.nl
Distributed in the USA & Canada by: A.A. Balkema Publishers, 2252 Ridge Road, Brookfield, VT 05036-9704, USA
Fax: 802.276.3837; info@ashgate.com
Printed in the Netherlands

Eurock '96, Barla (ed.) © 2000 Balkema, Rotterdam. ISBN 90 5410 843 6

Table of contents
Table des matières
Inhalt

Miscellaneous

Opening and closing session

Eurock '96, Barla (ed.) © 2000 Balkema, Rotterdam. ISBN 90 5410 843 6

Opening addresses

The Opening Ceremony of Eurock '96 took place at the "Sala 500" of the Lingotto Congress Centre. Lingotto is a new modern structure, specifically designed for Conventions, which is the result of the restructuring of the first FIAT - Lingotto car factory, built between 1917 and 1920.

Professor Barla (Chairman of the Scientific Committee):

Dear Guests, Ladies and Gentlemen

On behalf of the International Society for Rock Mechanics, the Eurock Committee, and the Eurock '96 Scientific Committee, it is a pleasure and a great honour to welcome you in Torino, Italy and Eurock '96, the 1996 International Symposium of the International Society for Rock Mechanics. This year, the International Symposium is on "Prediction and Performance in Rock Mechanics and Rock Engineering".

I have the privilege to introduce you Professor Bresso, President of the Provincia di Torino. I would like to mention that Professor Bresso is to be thanked for the marvellous welcome reception, yesterday evening at Palazzo Cisterna, in the city.

⇒ The head table (from left to right)
    Prof. G. Barla (Chairman of the Scientific Committee),
    Prof. M. Bresso (President of the Provincia di Torino),
    Prof. R. Zich (Rector of the Politecnico di Torino),
    Dr. S. Martinetti (President of the Italian Geotechnical Society),
    Prof. S. Sakurai (President of ISRM),
    Prof. M. Jamiolkowski (President of ISSMFE).

Welcome speech by Professor Mercedes Bresso
President of the Provincia di Torino

Voglio innanzitutto ringraziarVi anch'io per questo benvenuto. Sono stata lieta, ieri, di incontrare alcuni di voi, in occasione del drink di benvenuto.

Questa mattina invece l'occasione è molto più seria ed importante. L'augurio quindi di "Benvenuti e Buon Lavoro" che Vi rivolgo, a nome dell'Amministrazione della Provincia di Torino, non è formale perché, come molti di voi sanno ed avranno modo di vedere nelle numerose visite che ho visto nel programma, la Provincia di Torino è molto interessata, per la sua struttura fisica ma anche per la sua storia, ad un tema come quello della Meccanica delle Rocce. La nostra Provincia è in larghissima parte alpina e quindi ha grandi legami con il mondo delle rocce ed una storia anche di passione.

Le montagne sono per noi un oggetto continuo di lavoro. Vedrete molti dei più importanti trafori ferroviari ed autostradali che si siano fatti nel nostro Paese. Sapete che nella nostra Provincia è stato aperto il primo traforo ferroviario delle Alpi, quello del Frejus, e probabilmente sarà aperto il primo traforo alpino dell'alta velocità, l'Alpetunnel, che sarà il più lungo traforo ferroviaria delle Alpi, una volta realizzato.

E' quindi con particolare piacere che Vi do il benvenuto per questi lavori che hanno per noi molta rilevanza. Spero tuttavia che avrete anche l'occasione di visitare Torino e la sua Provincia, che offrono panorami naturali stupendi, un sistema di parchi nazionali, regionali e provinciali, unico in Europa, ed anche l'antica capitale d'Italia, una città che non solo ha palazzi e monumenti di grandissimo pregio e grande interesse, ma che ha un impianto urbano di grande piacevolezza, bellezza ed attrattività.

Vi auguro quindi buon lavoro ed anche, proprio perché siamo interessati a che riconosciate ed abbiate modo di apprezzare le qualità del nostro territorio, buon divertimento nei momenti liberi. Grazie.

Many thanks for your kind welcome. I have been very happy, yesterday, to meet some of you, on the occasion of the welcome drink.

Today, the occasion is much more serious and important. My wish for "welcome and good work" to you, on behalf of the *Amministrazione della Provincia di Torino*, is not formal. As many of you know and you will be able to see during your visits, the Province of *Torino* is highly interested, due to the physical structure of its territory and for its history, to the subject of Rock Mechanics. Our Province is alpine in large extent. The need is to deal with problems in the world of rocks. In a few occasions our relationship with rocks is a really true history of passion.

The mountains are for us a subject for continuous work. You will be able to see some of the most important railway and road tunnels constructed in our country. You know that the first alpine tunnel through the Alps, the Frejus tunnel, was excavated in our Province. Probably, also the first high speed alpine tunnel, the Alpetunnel, will be excavated in our Province and will be the longest one, once completed.

It is therefore a pleasure to wish you a great success in your work, which is of great interest for us. However, I hope that you will be able to visit *Torino* and its Province. They offer: marvelous sites, a system of national, regional and provincial parks, which is unique in Europe, and our city, the first capital of Italy, a city which is rich not only for buildings and monuments of great value and interest, but has also a urban planning of great beauty and attractiveness.

I wish you a good work, I hope that you will also be able to enjoy yourself during your free-time, as we are interested in your seeing our territory, so that you can appreciate it for its high quality. Thank you.

Address by Professor Rodolfo Zich
Rector of the Politecnico di Torino

Ladies and Gentlemen,

On behalf of the *Politecnico di Torino*, I cordially welcome you, as participants to Eurock '96, the 1996 International Symposium of ISRM, the International Society for Rock Mechanics.

We are honoured that more than 350 experts from 30 nations have come to *Torino*, which has been chosen by ISRM to host this International Symposium.

Eurock '96 has been organised by AGI, the Italian Geotechnical Society - National Group of ISRM, in cooperation with the Department of Structural Engineering and its Rock Mechanics and Rock Engineering Group, lead by Professor Barla.

The cooperation between a specialised Society and a University Department is a remarkable example of our attempt to promote exchanges between the outside world and our University. In the Italian University system, our school plays a unique role. The educational programs and research activities, with the tradition of more than 100 years in the various engineering fields, are well in balance with our continuous commitment to promote scientific research. It is of relevance here to mention that the *Politecnico* is the only institution in the entire Italian University system to rise 50 percent of its research funds through projects financed through external contracts.

As we approach the year 2000, Italy and Europe are moving toward the development of modern infrastructures such as railway links within cities, highways, high-speed railways, with some deep tunnels ranging in length up to 50 km. This takes place, without neglecting other important fields of endeavour, such as those dealing with the environment. All of this poses challenging tasks to Rock Engineering, and we expect our school to play a remarkable role in this respect.

Eurock '96 offers the occasion for a fruitful exchange of new ideas. With this in mind, I wish your Symposium a good success and all the participants a nice stay in *Torino*. Thank you.

Address by Dr. Eng. Sandro Martinetti
President of AGI Italian Geotechnical Society

Ladies and Gentlemen,

On behalf of the *Associazione Geotecnica Italiana*, the Italian National Group of ISRM, I take great pleasure in welcoming you to Turin and to Eurock '96, the 1996 ISRM International Symposium.

As chairman of the Organising Committee, I hope that we succeeded in creating the proper framework to assure a smooth and comfortable development of four days of intense and stimulating discussions on some fundamental aspects of modern Rock Mechanics and Rock Engineering.

I wish you a fruitful work and a pleasant stay in this noble city and for many aspects unknown city of Turin. Thank you for your attention.

**Address by Professor
Shun Sakurai**
President of ISRM

Mr. Chairman, Distinguished Guests, Ladies and Gentlemen,

It is my great honour and great pleasure to have this opportunity of giving a few words at the opening session of this Symposium.

First of all, on behalf of ISRM, I would like to welcome all of you to the 1996 International Symposium. I would like to congratulate the Organising Committee, the Italian Geotechnical Society, the Italian National Group of ISRM, and the Department of Structural Engineering of the Technical University of Turin for the excellent organisation of this Symposium.

The theme of this Symposium is "Prediction and Performance in Rock Mechanics and Rock Engineering", which is covering a very wide area of Rock Mechanics and Rock Engineering, from fundamental theory to practice. Actually, prediction of the behaviour of in situ rock masses is still a very difficult task, although many very nice fine computer programs have already been well developed.

This difficulty of course is mainly due to the fact that many uncertainties are involved in modelling rocks and in evaluating input data necessary for the prediction analysis. To overcome this difficulty, collaboration among scientists, researchers and practicing engineers becomes more and more important.

Considering the aim of the Symposium, I am strongly convinced that the Symposium can provide the forum for both researchers and practicing engineers to exchange ideas and opinions on various issues of Rock Mechanics and Rock Engineering, and as a result it should be possible to feel the gap between theory and practice.

Last but not least I would like to thank the Italian colleagues for hosting the 1996 ISRM International Symposium here in Turin, and for providing us with excellent facilities for ISRM Board, Council and Commissions meetings. These facilities have made the meetings very efficient and comfortable.

I wish the Symposium every success in the next four days. Thank you very much.

Address by Professor
Michele Jamiolkowski
President of ISSMFE

Mr. Chairman or better dear Giovanni, Distinguished Guests,
Ladies and Gentlemen, dear Colleagues

It is really a great pleasure and privilege to greet, on behalf of the International Society for Soil Mechanics and Foundation Engineering, the Participants to Eurock '96.

I am particularly pleased that such an important event is held in *Torino* and it has been organised by the Rock Mechanics group of the Technical University of *Torino*, which is also, by chance, my University.

The group headed by Professor Giovanni Barla has to be congratulated not only for taking on the organisation of this significant conference at a difficult time like this, but also for setting up a technical programme of remarkable level and great technical interest.

This conference has offered to the three presidents of the sister societies the favourable occasion to meet with our Australian friends and thus activate the common effort aimed at the organisation of the joint congress planned in Melbourne in the late fall of the year 2000.

The congress, named GeoEng2000, will deal with the topics of common interest for researchers and practitioners, members of the International Association of Engineering Geology, International Society for Rock Mechanics, and International Society for Soil Mechanics and Foundation Engineering.

It is aimed at strengthening the present achievements of the geomechanical disciplines as well as fostering further developments of the new scenarios that will open the eve of the third millenium. I thrust that GeoEng2000 may give us the occasion for a closer cooperation among the International Society of Engineering Geology, the International Society for Rock Mechanics, and the International Society for Soil Mechanics and Foundation Engineering. That hopefully can be regarded as an effort to explore the feasibility for the three societies to merge in one strong International Society for Geomechanics.

I wish all the participants a rewarding conference and an enjoyable stay in *Torino*. Thank you.

Eurock '96, Barla (ed.) © 2000 Balkema, Rotterdam. ISBN 90 5410 843 6

An overview of the Eurock '96 Technical Program

Giovanni Barla – *Chairman of the Scientific Committee*

Eurock '96 is a Symposium of the Eurock (Europe + Rock!) series, which was initiated in 1992 with the intent to provide a forum for discussion on Rock Mechanics and Rock Engineering among participants from European Countries. However the meetings became soon globally international, as it is well demonstrated by the present ISRM International Symposium, which sees in *Torino* more than 350 participants from 35 different Countries.

The intent of Eurock '96 is to discuss the "Design and Performance" theme, which was addressed for the first time in Europe, in the conference held in Cambridge in 1984. However, the theme will be dealt more extensively in this Symposium. Fundamental aspects of Rock Mechanics and Rock Engineering will be discussed, in conjunction with the use of analytical and computational methods. Also to be considered is the adoption of observation and monitoring programs, in the framework of case histories. In all cases, the intention is to compare prediction with observed behaviour and performance.

The motivations to address at this stage the prediction and performance theme in the context of Rock Mechanics and Rock Engineering is the need to question ourselves, on the present level of predictive capability, when facing the solution of a rock engineering problem.

The prediction of rock behaviour at the laboratory and in situ scale, ... the prediction of the performance of rock structures to be built in and within rock, ... are formidable tasks in Engineering Rock Mechanics. It is my firm opinion that our ability to show to others that our methods are capable to provide reliable predictions in engineering practice is the most formidable challenge for the years to come.

I would like to say now a few words on the organisation of this Symposium. We have received a total of 320 summaries , which were reviewed by the Scientific Committee to come down to a total of 178 papers. Out of these initial proposals, a total of 178 papers were published in the two volumes of Proceedings, which were distributed this morning. For your information it is of interest to compare the two slides below, where the distribution of participants to Eurock '96 is compared with the distribution of technical papers.

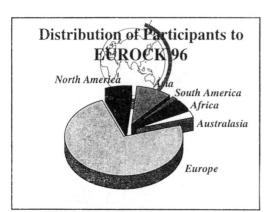

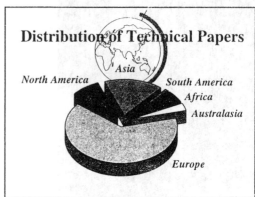

As you will be able to see from the Final Program, the Symposium comprises a number of Special Lectures to be given in Plenary Sessions. Among them, I would like to mention the lecture by the Rocha and the Schlumberger Award winners.

Special Lectures
(PLENARY SESSIONS)

+ *ROCHA MEDAL LECTURE*
+ *SCHLUMBERGER LECTURE AWARD 1996*
+ *KEY-NOTE LECTURES*
+ *INVITED LECTURES*
+ *GENERAL REPORTS*

Workshops

+ **Rock mass modelling: continuum versus discontinuum.**
+ **Rock mechanics modelling of large scale natural phenomena.**
+ **Fluid-rock interaction in saturated and partially saturated media.**
+ **Risk assessment in rock engineering.**

A special interest of this Symposium was posed on the organisation of four workshops.

WK1   is the workshop on "Rock Mass modelling"

Numerical modelling is widely used today for the solution of rock engineering problems, both in fundamental studies and engineering applications. The existence of discontinuities at all scales in geological materials has led to the development of discontinuum modelling methods to supplement more traditional continuum approaches, both for fluid flow and stress distribution problems.

In a number of cases one faces the choice of the model to adopt for the simulation of rock mass behaviour, i.e.: should a continuum or a discontinuum modelling approach be used? This and other topics will be discussed during this workshop, with the purpose to address pending issues, future lines of research, and trends in present engineering practice.

WK2   is the workshop on "Rock mechanics modelling of large scale natural phenomena"

Some of the key issues associated with rock mechanics aspects of large scale natural phenomena such as rapid mass movements (debris flows or avalanches) and creep deformation of mountain slopes will be discussed. Rock mechanics studies of these events may be carried out by physical modelling, numerical analyses or by empirical observations.

The workshop will consider the present state of the art of each of these approaches. These large scale phenomena differ from the smaller scale events on a normal engineering scale since the failure mechanisms and governing parameters are generally different and not well understood, and site investigation data is generally sparse by comparison with the size of the problem. Further difficulties arise as a result of extremely slow or rapid movements where creep or dynamic effects need to be taken into account.

WK3   is the workshop on "Fluid-rock interaction in saturated and partially saturated media"

The coupling between pore pressure, temperature and mechanical response of porous media has been the object of numerous developments over the past ten years and it can be stated that for conventional porous rocks, the industry possesses a complete formal framework with which to study field cases, i.e. thermo-poro-mechanics. Nevertheless, as is the case with every theory, limits of applications have been found, among which the proponents of this workshop point out the following: the cases where dissolution/recrystalization takes place at grain contacts; the cases where the rock is saturated by two-phase fluids (oil/water, air/water); the cases of shales where electrical forces are known to bound the water to the rock skeleton; the case of chalk where a sudden collapse may occur upon imbibition. All of these topics, which have many industrial applications, i.e. in the oil industry, tunnelling, civil engineering, radioactive waste storage, etc. are today debated among researchers and engineers. The aim of the workshop will therefore be to review these pending issues and to delimitate the most promising roads towards a rational resolution.

WK4   is the workshop on "Risk assessment in rock engineering"

This workshop will be in form of debates on three areas of rock engineering, namely tunnels, slopes and flows. In each case there will be a motion on the table; a proponent, an opponent, a second for, and a second against will provide brief prepared arguments followed by an open discussion involving all workshop participants. At the end of the workshop there will be a general discussion. The debate format has been proposed to help focus the arguments and also to promote a lively discussion. The debaters have experience with probabilistic methods and risk analysis.

Finally, the technical program moves to the Plenary Sessions and Oral presentations of 60 papers, in addition to 118 posters. Overall the main themes and sessions are briefly summarised in the table shown below:

Session	Theme	Title	No. Oral presentations	No. posters
1 A - 1B	1	Fundamental Aspects in Rock Mechanics and Rock Engineering	20	40
2	2	Near Surface Rock Engineering	10	16
3 A - 3 B	3	Rock Engineering at Depth	20	43
4	4	Environmental Rock Engineering	10	19

Before proceeding any further, I would like to say a few words on the framework and organisation of the various sessions. You will notice, through the proceedings, that I have taken the liberty to subdivide each session in subthemes. Obviously, by doing so, I have used some liberty, but I may also have made some mistakes or taken decisions which are different from your point of view or your desire.

It is however my hope that, by having tried to give a logic and orderly sequence to the various contributions received, we will be able to question ourselves on our present capability to make appropriate predictions in Rock Mechanics and Rock Engineering. I would also like to ask for your help in developing interesting and fruitful discussions throughout Eurock '96. Thank you for your attention.

Eurock '96, Barla (ed.) © 2000 Balkema, Rotterdam. ISBN 90 5410 843 6

Closing addresses

Professor Barla:

Ladies and Gentlemen, dear Colleagues,

It is my duty to report briefly on Eurock '96. A few figures, first: we had a total of 360 registrants and 40 accompanying persons, 14 students from the Politecnico di Torino, and more than 40 persons attending from the Technical Exibition, in addition to the 10 invited persons who were present during the opening session. As I showed during the first day of our Symposium, the distribution of technical papers and participants to Eurock '96 is similar. We had a wide participation from Europe, however all the continents were well represented.

We started on Saturday, with the ISRM Board, to move to the Commissions and Council meetings, which took place on Sunday. I would like to report that a total of 35 National Groups were present during the Council meeting. Also important is to mention that this meeting was attended by the Presidents of both ISSMFE and IAEG, the sister societies, and by the President of ITA, the International Tunnelling Association. This is definitely a case record for Eurock '96!

The closing session took place with speeches by Prof. Barla, Prof. Kim, the Organiser of the 1997 ISRM International Symposium in New York, and Prof. Sakurai, who declared Eurock '96 closed (*right*)

The Symposium was initiated on Monday morning and moved smoothly through a number of important technical sessions, starting with the outstanding presentations of Dr. Board, the Rocha Medal 1996 winner, and Mr. Windsor, the Schlumberger Award winner, to move to a stimulating lecture by Professor Goodman. I would like to mention here the high quality of the key-note lectures by Dr. Nova, Sharp, Maury and Dershowitz, and of the general reports.

As you know, we have organised four workshops which took place concurrently on Monday afternoon. I would like to say a few words on the technical contents of these workshops, based on the information received by the Chairmen. In a number of cases, the discussion during the workshops was on papers published in the proceedings. In other cases, new contributions and ongoing works were presented.

⇒ The Workshop 1 on "Continuum versus Discontinuum modelling" was intended to examine the realm of applications of continuum and discrete numerical models in Rock Mechanics. The contributions examined a range of applications, primarily concerned with jointed or blocky rock and considered both solid deformation questions and the ability to represent two-dimensional fluid flow in jointed rock by a continuum probability approach.

With respect to the representation of discrete solid blocks and discontinuities, it was agreed that it is possible to use equivalent anisotropic continuum representations of systems of ordered discontinuities, within limits. As a rough approximation, it was agreed that, when the characteristic size of the discontinuity was less than 0.1 to 0.2 of the scale of the significant geometrical parameters in the problem, as an example the diameter of an excavation, and assuming that all the discontinuities exhibited identical properties - which may not be realistic - than an equivalent continuum analysis is acceptable. Overall it was agreed that there was still a need for both continuum and discontinuum models. Each can often be used to solve the same problem correctly, but there are applications where each method has advantages. For fluid flow, it appears however that there is as yet not satisfactory equivalent continuum representation of a discretely fractured rock mass.

⇒ The Workshop 2 on "Rock Mechanics modelling of large scale natural phenomena" focused on two particular types of large scale natural phenomena which are of particular importance to Rock Engineering: these are large scale rapid mass movements and gravitational creep deformations of mountain slopes. Large phenomena were accepted as being those with a lower limit of 1 million cubic meters.

These large scale natural phenomena pose formidably difficult subjects for modelling because of the huge differences in scale, mechanisms and processes compared with the events on a more conventional engineering scale. Problems of scale involve two different aspects. Firstly the size and complexity of the events mean that site investigation data are generally sparse or non-existent and the geological mechanism is therefore inadequately defined. Small scale laboratory or in situ tests may give little guidance to the parameters controlling the processes. Scale effects of these orders of magnitudes have not been extensively studied in a systematic manner. Further difficulties arise as a result of extremely slow or rapid movements where creep or dynamic effects need to be taken into account.

The workshop discussions acknowledged: a) the fundamental problem of defining mechanisms and parameters for these very large events; b) the general economic impracticability of carrying out exploration and investigation to normal engineering standards, and c) the general impossibility of providing engineering remedies for these events.

Despite these serious handicaps, there has been some significant progress in research in these areas over the last ten years. The observational database has been extended to provide more information on the geological mechanisms involved, and this must continue to be a key priority for further research. Modelling of these events has tended to develop along two separate lines depending on the rate of the event: creep events have typically been modelled by numerical methods with parameter selection being fairly subjective and calibration against field situations being essential; rapid events have generally been modelled by physical means, although these are developments combining field studies together with physical and numerical modelling.

There are clear pragmatic reasons for modelling having developed in the above way. However, there seemed to be a consensus that modelling of such events should utilise as far as practicable all the tools in the rock engineers' armoury and not place undue reliance on one or other approach, at least until reliable methods capable of accurate prediction have been developed.

⇒ The workshop on "Fluid/Rock Interaction in saturated and unsaturated porous media" has allowed to measure the important progress made in that field since 1989. In particular, it is now clear that: a) the introduction of a second phase in a rock with small pores may have dramatic effects on its behaviour because of capillary effects; b) in the case of collapsible rocks, complete rheological models have now been developed accounting for most experimental observations but real industrial use of such models remains in its infancy; c) the chemical interaction between salted brines and rocks such as shales have been proved to also change significantly their behaviour.

⇒ Finally, a short comment on workshop 4, which was on "Risk assessment in Rock Engineering". As intended, the debate format allowed the workshop participants to focus on important and controversial issues.
The debates and the discussers from the audience covered most of what is important in risk assessment with reference to rock tunnelling, rock slopes and fluid flow through fractured rock. I was told that many of you found it particularly useful, as a number of practical applications were discussed, which put risk assessment into the context of Rock Engineering.

We had, as you know, eight technical sessions with the oral presentation of 60 papers and poster sessions which were well attended. The technical sessions covered the main themes of the Symposium, which were subdivided in sub-themes. For the main themes - Fundamental aspects in rock mechanics and rock engineering, Near surface rock engineering, Rock Engineering at depth, and Environmental Engineering we had excellent general reports by Dr. Kaiser, Egger, Duddeck and Stephansson. As a summary we can say that the papers presented reflect a broad variety of problems encountered in theory and practice. They are as well a good representation of the state of the art and in cases show promising and innovative procedures.

As ISRM Vice-President for Europe and Chairman of the Eurock Committee it is my privilege now to report on the Eurock Symposia planned for the near future. Eurock '97 will take place in Vienna as a joint ITA-ISRM conference. In order to stress the role of Rock Mechanics and

Rock Engineering in Tunnelling, at the design and construction stages, a workshop on Rock Mass Characterisation is being organised in conjunction with the main conference activities. Eurock '98 will see once again the joint SPE-ISRM organisation of a symposium on Petroleum Engineering, which is to take place in Trondheim on July 1998. In both cases, we will have the opportunity to stress the importance of a multidisciplinary approach to Rock Mechanics. We are also looking ahead to the new millenium and Eurock '2000. I am pleased to announce in this respect that we have three outstanding proposals.

It is my duty now to acknowledge the help of our Sponsors, the support of Owners and Contractors, who have assisted us in making the six technical visits very successful and the Exhibitors who helped in putting together the technical exhibition of rock mechanics instrumentation, rock engineering equipment and technology and contributed to the success of our Symposium.

Finally allow me to extend my appreciation and gratitude to the many people who have been working hard to make Eurock '96 a successful event: the key-note and invited lecturers, general reporters, authors, chairmen and co-chairmen. I would also like to thank the interpreters, who have assisted us so nicely and efficiently and our hostesses here, at the Lingotto Center, as well as during the technical visits.

I would like to extend my heartful thanks to Dr. Martinetti, the President of the Italian Geotechnical Society, who chaired the Organising Committee. I would like to thank him for the continuous support from the initial proposal to the very end, independent from the difficulties encountered in a somewhat hard time for Rock Engineering and Geotechnical Engineering in Italy today.

My thanks should go to the Master and Doctoral students of the Politecnico di Torino, who have tried their best in making your way comfortable; they were reporting to me continuously during the Symposium: in this way we solved the few problems, which were coming up some time, quickly and in a reasonable manner. I would also like to thank Elena and Susanna from AGI, who worked efficiently in the Secretariat on behalf of Eurock '96.

It is my pleasure to acknowledge as well the help of Dr. Marco Barla and Dr. Santina Aiassa for the splendid work they have done in getting all the slides organized and everything going very efficiently. Last and not least, I would like to thank Ms. Simona Verdun and Ms. Olivia Montaruli, whom we owe a special word of thanks. They have been working with me in at least the last two-years.

Finally, I would like to thank my wife Bruna. I owe her a lot for the many hours spent together, days and nights, in getting the many things going ahead, in particular the correspondence with many of you. Thank you very much.

Address by Professor Kim

Thank you Professor Barla. I like to extend my heartfelt congratulations to you, Professor Barla, and the organisers of this Symposium. I really enjoyed every moment since I arrived in Torino. I enjoyed the dinner at the Castello. I enjoyed the concert last night. You really established a new standard on conducting ISRM symposia.

I feel very pressed to have a wonderful symposium next year in New York city. New York city, as you know, is the greatest city in the world. We pose ourselves as the capital of the world. Of course, the technical content of the program is going to be the most important thing. So, our Community will work hard from now until next year to create a wonderful technical program. We cannot build the castle in one year, but we can build a really good technical program, I thrust. Of course we need your help. I encourage you to submit abstracts through internet or by fax, or by regular mail by October 1st, and we will respond to you very soon.

Of course we will have a wonderful program for accompanying persons. We will visit many wonderful places in New York city: the Tiffany's store on the Fifth Avenue, the Metropolitan Museum of Art, ... Anyway, those of you who have not visited New York city are invited to visit my web page first, through my internet. We have all the information on hundreds of Broadway's shows going on, the Metropolitan Opera, and many other cultural activities. You can down-load subways maps. I will add more information in the next several months and also we will be very responsive to your need.

I was told that next year is going to be a very important year for ISRM, as Professor Sakurai will mention about three important issues. I like to invite again each one of you to the great city of New York, to Columbia University, next year. Remember that the Symposium will be held just a few days before the Independence Day, July 1st. The city itself will have a big celebration, so it will not be very boring while you are staying in New York city. Thank you very much, and I love to have all of you in New York.

Address by Professor Sakurai

Thank you Professor Barla, Ladies and Gentlemen, Colleagues,

The Symposium is now approaching to the very end. I am sure all of you are becoming tired, after seating four days in this conference room.

I hope that you have enjoyed your participation in this Symposium, not only in technical but also in social events. We have no doubt that the Symposium has been of high quality and a great success, as Professor Barla has now described in detail. On behalf of ISRM I would like to congratulate Professor Barla and the Organising Committee for a very successful Symposium. I would like to thank Dr. and Mrs. Barla for providing us a very kind hospitality in the various occasions.

Now I would like to make a few general remarks on this Symposium. Many excellent papers have been presented and many valuable discussions have been taking place. But I have the impression that there is some disconnection between theoretical papers and practical papers. As far as my impression is concerned, both seem to be rather independent from each other. This fact indicates that we still have a question on how to apply theory to practice. We have discussed this issue for quite a long time and many times.

In order to answer this question, I am convinced that we have to see more actual behaviour of rock and rock structures. To do this we need more case histories which can provide us with chances to know what happens in the field. As a result we can prove how accurately our theory can be applied to practice. In this respect we should remind that computer programs can represent the behaviour of rock and rock structures, but the actual behaviour cannot represent the computer result. Therefore I think we have to pay more attention to what is going in the field. Anyway, we should make a continuous effort on how to apply theory to practice and how to fill the gap between theory and practice.

As Professor Kim has already mentioned, the next International ISRM Symposium will be held in New York in June next year. This Symposium is very important not only from the technical point of view but also from our Rock Mechanics Society point of view, as we have to take significant decisions. We have three important issues: one is the President elect, who is to take responsibility for the next term of office, starting from 1999 to 2003; then, we have to choose the next ISRM Congress Venue, after Paris, and the third one the name change of our Society.

The ISRM Board is proposing the name change of our Society. I have already explained in the Council Meeting why we have to change. Probably I can simply explain why we have to change at this stage the name. As I mentioned earlier, we need more chances to look into practice, that means we have to get more practicing engineers to pay more attention to our Society. That is the reason why we are now proposing the new name to be "International Society for Rock Mechanics and Rock Engineering". In the next Symposium in New York we have to take a decision. Once again I would like to thank Professor Barla and his colleagues for their warm hospitality. I would also like to thank the interpreters and the ladies working in this conference room for their contribution.

Now, I am pleased to announce that the 1996 ISRM International Symposium is closed. I wish all of you a pleasant journey back home, and see you again in New York. Thank you.

Special lectures

Dr. Mark Board
Rocha Medal Winner

Mr. Christopher Windsor
Schlumberger Prize Winner

Prof. Richard Goodman

Eurock '96, Barla (ed.) © 2000 Balkema, Rotterdam, ISBN 90 5809 843 6

Numerical examination of mining-induced seismicity

Mark Board

Itasca Consulting Group Incorporated, Minneapolis, Minn., USA

ABSTRACT: This paper describes a numerical implementation of the Excess Shear Stress (ESS) approach proposed by Ryder (1987) for providing an engineering estimate of the seismic potential of a rock mass. An approach is presented in which a geologic model of the structure of a rock mass is first developed using standard mapping techniques. A separate numerical analysis is conducted to estimate the mining-induced stress state for a given mining sequence. The stress state at each mining step is superimposed on the fracture model, and the ESS method is used to derive an estimate of the resulting seismic source terms. These source terms are used in comparison to the seismic data base provided by a waveform digitizing system as a means of calibration of the model. Once validated, the model can be used for relative comparisons of various methods or stope sequencing for mining a block of ground. An example application for examination of seismicity at the Lucky Friday Mine, Idaho, USA, is presented.

1 INTRODUCTION

In the future, ever-greater reliance will be made on the extraction of deep ore reserves. Already, mining in South African gold mines exceeds 3500 m, and is approaching 3000 m in many mining districts around the world. One of the great technical challenges and one of the primary safety and production problems associated with deep mining is rockbursting. Rock-bursting is the violent failure of a rock mass in close proximity to an underground excavation, often resulting in the expulsion of rock fragments into the opening at high velocities. The primary modes of rockbursting generally recognized in the literature are instabilities of small volumes of rock associated with localized stress concentrations at the mining face (termed crush or volume bursts) as well as shear-failure within the rock mass, either on pre-existing geologic discontinuities or through formation of new shearing fractures. The two approaches which are commonly used in the mining industry in dealing with rockbursting include: (1) attempts at real-time prediction using the data produced by in-mine seismic monitoring systems; and (2) mine planning through practical experience, or through the use of numerical representations of seismic source mechanisms. This paper addresses the latter approach and presents a methodology for employing numerical methods for representation of seismicity related to the predominant shear-failure mode.

The problem of mine seismicity is, in general, stochastic in nature due to the variability in distribution (e.g., fracture lengths and orientation) of geologic structure in the rock mass. In some cases, the controlling geologic structures are well known, as in the case of slip on large-scale faults. In other cases, the seismicity is related to families of through-going joint or bedding structures, shear zones or faults whose geometry can be described statistically, but is not known with great certainty. A methodology for assessing the seismic potential of a rock mass is presented which recognizes both the stochastic nature of mine seismicity as well as the deterministic nature associated with faults whose seismic behavior is well-known to the mine staff.

In order to represent seismicity associated with shear failure on existing discontinuities, the problem of representation of the stochastic nature of rock mass fracturing in a numerical method must first be addressed. Using work from the literature, a scheme is presented for random generation of a "representative" rock mass using probability density functions of the rock mass fracture geometry as derived from underground detailed line mapping. In this case, the number of possible fractures to be represented is manageable since we are concerned only with those whose continuous length is great enough to generate seismic events.

It is impossible to explicitly examine the slip potential of all possible fractures, so an approximate method for estimation of seismic potential is developed, based on work by Salamon (1993). The mining-induced stress state, determined using a separate numerical analysis, is mapped onto the location and orientation of each of the statistically-defined structural features. The slip potential for each fracture is determined using the standard Mohr-Coulomb slip condition, and an estimate of the resulting slip area, ride and approximate seismic moment determined using the Excess Shear Stress (ESS) technique of Ryder (1987). Using modem personal computers, it is possible to examine the slip potential and approximate seismic output of a very large number of fractures, thus obtaining a statistical representation of the seismic potential of a complex rock mass. More importantly, the effect of the geometric and strength properties of fractures on their seismic potential can be examined efficiently, thereby allowing the engineer to define possible seismic source mechanisms.

For such a model to be useful, the results must be verified against field data. Fortunately, most mines which experience a rockbursting problem have developed an extensive seismological data base which often includes estimates of source parameters as well as the source locations. Since the numerical model is capable of producing the equivalent source information, it is often possible to determine the correspondence between the predicted and actual seismic behavior. This, in turn, leads to a better understanding of the geological and mining factors which control the seismicity. It is suggested that the model can then be used for relative assessments of the seismic potential of varying mining methods or sequences, as opposed to attempts at precise predictions of seismic response. In this way, mining methods which minimize the dominant source mechanism can be specified.

A demonstration of the method is given for a parameter study of the seismicity associated with advance of a thin, horizontal longwall stope at great depth. The impact of the variability of the geometric properties and frictional strength of the geologic structures as well as the in situ stress state on seismic response are examined through the use of standard Gutenburg-Richter plots of event frequency and magnitude. An application of the model to analysis of the seismicity at the Lucky Friday Mine

in Idaho is given. The model is used to elucidate the mechanisms for the rockbursting, and practical conclusions regarding mining methods which will reduce seismicity are presented.

2 MECHANISMS OF ROCKBURSTING

2.1 Evidence for predominance of shear-failure as a rockburst mechanism

There appears to be general agreement that the mechanisms for most rockbursting fall into two classes (Ryder, 1987): one associated with the crushing of highly stressed volumes of rock, and one associated with unstable slip or rupture along weakness planes in the rock mass. The first class is often referred to as "strain," "crush" or "volume" bursting, and the second as "fault-slip" rockbursting.

Gibowicz (1990) relates the strain - or volume-failure events directly to the geometry of mining operations. Volume events tend to occur where the mine geometry results in stress concentrations, such as in pillars or at an advancing mining face. The fault-slip mode of rockbursting may occur in close proximity to the excavations (Hedley et al., 1985) or may occur hundreds of meters within the rock mass. In either case, as pointed out by Gibowicz (1990, p. 4):
- mine seismicity is strongly affected by local geology and tectonics, i.e., by rock mass
- inhomogeneities and discontinuities, and interactions among mining, lithostatic, and residual tectonic stresses on local and regional scales.

Therefore, even though rockbursts are often grouped into the two convenient classes given above, geologic discontinuities and their interaction with the in situ and mining-induced stresses are of primary importance in control of the failure response of the rock mass and, as described below, are generally attributable to the same shear failure mechanism.

The advent of waveform-recording seismic systems in the late 1970s has allowed detailed studies of the source mechanisms of seismic tremors (e.g., Gibowicz, 1990). A significant body of data has now been published which identifies a primary shear failure mechanism for mine seismicity, in which the failure is associated with pre-existing planes of geologic structure. Typical of the field studies is the work of Spottiswoode (1984), who analyzed source mechanisms from 11 seismic events at a South African gold mine. His interpretation was that the mechanisms of the events were best-fit assuming a pure shear source with no volume change. Fault-plane solutions showed that the events were attributed to shear failure on planes striking parallel to the face direction or to dykes intersecting the face. These conclusions are consistent with underground observations of fractures observed in the stope hangingwall.

2.2 *Mechanical model for representation of shear failure on geologic discontinuities*

Mining engineers often regard fault planes as planar features with smooth surfaces; in fact, Scholz (1990) shows that faults have roughness at all length scales. This roughness may take the form of topographical variations on the surface such as striations, and both microscopic and macroscopic waviness and splaying of the fault from a main branch into multiple features. These and other forms of macroscopic asperities, including intersections of faults with dykes or offsets by other structures, can create regions of high cohesive strength on the fault surface. Ryder (1987) examined the problem of an excavation in a high vertical to horizontal stress field in a fractured rock mass (Figure 1).

As the regions of increased shearing stress above and below the face "sweep" through the rock mass with the progress of mining, slip may occur on structures influenced by the increased shears.

Occasionally, a long continuous structure such as a fault may be affected. Slip on the fault is possible if the stress state satisfies the Mohr-Coulomb slip condition:

$$\tau_s = c + \mu_s \sigma_n \qquad (1)$$

where μ_s = static shear strength,
μ_s = static friction coefficient,
σ_N = normal stress, and
c = cohesive strength.

The cohesion, c, along the discontinuity is governed by infilling or perhaps the degree of "welding" or adhesion of asperities. Once the static strength is overcome and slip initiates, dynamic conditions prevail, and the friction coefficient will drop suddenly to μ_d. Fault plane motion will then be resisted by the dynamic shear resistance (τ_d), where:

$$\tau_d = \mu_d \sigma_n \qquad (2)$$

Here, it is assumed that once dynamic conditions prevail, all cohesion has been eliminated along the slipping surface. The reduction in strength from a static to dynamic state can be viewed on a standard Mohr diagram and in terms of the shear stress-shear displacement behavior as shown in Figures 2a and 2 b. Under static conditions, the shear strength will be a function of the normal stress, cohesion and static friction. When peak strength is reached, the cohesion introduced by strength variations along the

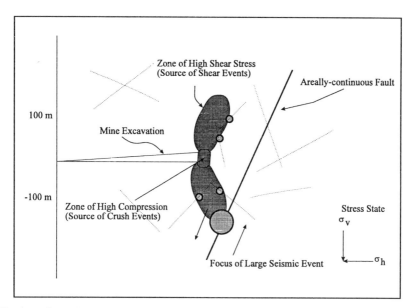

Figure 1. Cross-section taken through an advancing stope face illustrating the zones of high shear and compressive stresses around the face in a high vertical to horizontal stress ratio stress field.

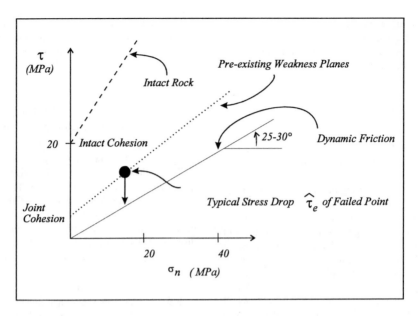

Figure 2a. Softening response of discontinuity from static to dynamic strength during slip as shown on a Mohr-Coulomb diagram [Ryder, 1987].

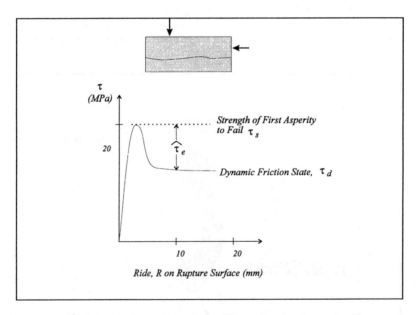

Figure 2b. shear stress-ride behavior for a discontinuity illustrating the "stress drop" occurring during slip [Ryder, 1987].

fault will be destroyed along the rupture, and strength will be reduced to a function of the normal stress and dynamic friction angle only. The peak stress-drop potential, termed $\hat{\tau}_c$ by Ryder (1987), is determined by the difference in shear strength of the two failure surfaces. This peak excess shear stress drop is equivalent to shear of the "most energetic asperity" (McGarr, 1981). Since the static friction angle and the cohesion of the fault surfaces may be highly variable, Ryder suggests that seismological evidence be used to estimate $\hat{\tau}_c$. A value of 5 to 10 MPa is suggested for unstable slip on planes of weakness, while 20 MPa (essentially the cohesive shear strength of the intact material) is suggested for unstable rupture of intact rock.

Figure 3 is a conceptual depiction of the variation of static strength on a fault. Significant local variability may result from asperities or strong regions, as seen in the previous figure. Superimposed on this plot is the variation in shear stress along the fault surface. This stress variation is a consequence of mining in the surrounding region. Figure 3 also includes a line to represent the dynamic shear strength of the fault, given by $\mu_d\sigma_n$. The stress field surrounding the excavation will be disrupted by the presence of the opening, resulting in stress concentration and reorientation ahead of the face. As the excavation approaches the fault, the shear stress on the fault surface may eventually

reach the static shear strength $(c + \mu_s\sigma_n)$ at some location, thereby initiating rupture. The stress drop accompanying shearing of this initiation point will have a maximum, $\hat{\tau}_c$, determined by the difference in the peak (static) and dynamic shear strengths at this point. To determine the ultimate region of slip which will occur after the initiation of rupture, one must examine the stress conditions on the surrounding fault surface. The excess shear stress (ESS) (τ_e) at any point along the fault is the magnitude of shearing stress in excess of the shear dynamic frictional strength:

$$\tau_e = |\tau| - \mu_d\sigma_n \qquad (3)$$

where: τ_e = excess shear stress (ESS),

τ = shear stress acting on the fault, and

$\mu_d\sigma_n$ = shear stress due to dynamic friction.

Positive values of the ESS indicate that slip is possible under dynamic conditions. Thus, once rupture begins, it will propagate from the initiation point, possibly through other asperities, eventually stopping when the ESS is sufficiently negative to inhibit further movement. Variations in static strength along the fault are largely irrelevant since, once rupture begins, it is assumed that the dynamic strength is operative and continuation of rupture depends on the ESS distribution. A consequence of

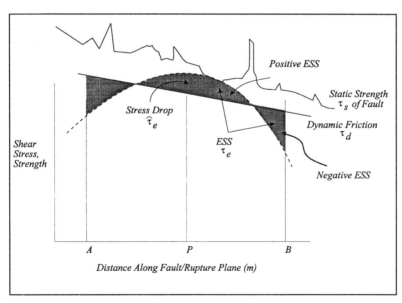

Figure 3. Conceptual depiction of the variation of shear strength and shear stress along a fault (The variation in shear strength may be a function of the topography of the fault or variations in in-filling materials.) [after Ryder, 1987].

1473

the assumption that the dynamic frictional strength along the fault is uniform is that the ride and seismic event magnitude may be over-predicted. This reinforces the need for verification of this approach against field data.

The stress drop $\Delta\tau$, and the length (L_s) of the fault subjected to positive ESS, can be used to estimate the average relative shear displacement of the opposing surfaces of the fault, with the analytical expression for a circular displacement dislocation [Salamon (1964)] in the following equation:

$$u_{ave} = \frac{8(1-v)\Delta\tau L_s}{3\pi(1-v/2)G} \qquad (4)$$

Once shear stress drop, radius (or area) of slip of the event, and relative shear displacements have been determined, it is possible to make estimates of the seismic source parameters of the ensuing event. The relevant parameter to most mining engineers is the event magnitude, expressed in a local magnitude scale. A robust measure of the magnitude of a shear event is the scalar magnitude of the seismic moment tensor, given by (Aki and Richards, 1980):

$$M_o = Gu_{ave}A \qquad (5)$$

where: M_o = seismic moment,
G = shear modulus of the rock mass,
u_{ave} = mean ride or slip vector averaged over the fault surface,
A = area of the slipping fault surface.

Once the seismic moment is estimated, scaling laws for converting moment to local magnitude can be used. Spottiswoode and McGarr (1975) established the following relation for local Richter magnitude, M_L, in terms of M_o (units of MN - m):

$$\log M_o = 1.2\, M_L + 4.7 \qquad (6)$$

3. IMPLEMENTATION OF TRE EXCESS SHEAR STRESS METHOD IN A NUMERICAL SCHEME

3.1 Introduction

The ESS approach has been applied successfully to back-analysis of numerous fault-slip seismic events in South African mines in particular, and has shown great promise as a tool for estimating the seismic potential of individual structures. General application of the ESS approach as a design tool faces a difficulty in that it is presumed that the engineer knows, in advance, the locations and orientations of critical failure planes. In this sense, the problem of estimation of fault- or discontinuity-related seismi-

city is, to some extent, deterministic in nature in that one often seeks to model the mechanical response of uniquely defined geologic structures. However, as pointed out by Morrison et al. (1993), the problem of discontinuity-related seismicity is often far more complicated due to the complex geologic structure of the typical rock mass. These authors discuss the need for ... development of some form of hybrid deterministic model, which incorporates aspects of weakly chaotic systems ... The object of this kind of simulation would be to generate the range of likely responses of the discontinuities rather than to calculate a unique solution to a particular set of pre-determined conditions (p. 236).

The main difficulty in modeling the seismic response of geologic discontinuities is the inherent uncertainty in the knowledge of rock mass structure. In this sense, the problem can be considered to be of the "data-limited" variety discussed by Starfield and Cundall (1988). Data-limited problems are defined as those for which little detailed information is available on either the structure or mechanical behavior of the material. This lack of data is typical of problems encountered in geotechnical and mining engineering. The approach suggested by these authors is exactly that expressed by Morrison et al. - that the model should be used as a tool to supplement the intuitive sense of the engineer - providing a means for checking hypotheses, or making parametric evaluations, rather than as a tool for making specific predictions or precise calculations.

A simple numerical approach to simulation of mine seismicity is suggested for analysis of the data-limited problem of the seismic response of a fractured rock mass. The technique employs a methodology similar to that suggested by Salamon (1993), which recognizes:

(1) the statistical nature of the distribution and geometry of the discontinuities which make up the "fabric" of the rock mass; and

(2) the fact that major faults, shears and dykes may occur sporadically and may not be particularly amenable to statistical description, although their geometries may be fairly well-established by the mine geologist.

A detailed knowledge of the mine geology is identified as a key parameter in the development of a seismic simulation method. The seismic source parameters for a given in situ stress state and mining geometry are estimated using the ESS method discussed previously. The unique aspect of the suggested approach is that an exact prediction of the timing and location of rockbursts is not attempted - this is felt to be a relatively fruitless approach in most circumstances. Instead, a methodology is developed for determining the most likely locations and magnitudes for seismicity based on the given degree of fracturing and mining-induced stress conditions. In this way, the modeling can be used to guide appropriate mine designs in rock-burst-prone conditions.

3.2 A seismic source simulation approach

Figure 4 presents a methodology for simulation of shear-related seismic source mechanisms using the ESS approach. A "representative" rock mass fracture network is first developed statistically for a given mine, based on a combination of underground detailed line mapping as well as geologic-scale mapping of major, through-going structures. [A simple numerical fracture generation scheme based on those used in modeling of fluid flow in fracture systems (e.g., Kulatilake, et. al., 1990) has been developed]. Detailed line mapping is then used to develop probability density functions of the spacing, dip and dip direction of the fractures which have continuous length in excess of a few meters. The length variation is assumed to fit a general negative exponential distribution (Priest and Hudson, 1981) with exponential decay coefficients determined from the field mapping. The geometry of the fracture network is generated, with random locations conforming to the

probability density functions. This network may be supplemented with the location, attitude and length of known fractures or structural features as desired.

In a separate step, a two- or three-dimensional stress analysis of the mining geometry for all steps in a mining sequence is performed. In all of the analyses conducted in the thesis, it was assumed that the rock mass behaved elastically. The type of numerical method used is unimportant, except that the method must be capable of determining the mining-induced stresses in a dense array of points in the rock mass surrounding the excavations. In a post-processing step, the calculated stresses are projected onto the locations of the fractures at many points on their surfaces. The shear and normal stresses are determined at each point on each fracture, and the static Mohr-Coulomb slip condition examined. If this criterion is exceeded for the assumed values of static friction angle and cohesive strength, slip is assumed and the area of positive ESS is determined. The resulting values of shear stress drop, area of slip and average ride (relative shear displacement) are used to estimate a seismic moment, and, through the local scaling relationship, a magnitude. Cycling through each fracture is a fairly rapid procedure using a personal computer, since only stress rotations are involved. Two assumptions are made based on field observation (Salamon, 1993), thus:

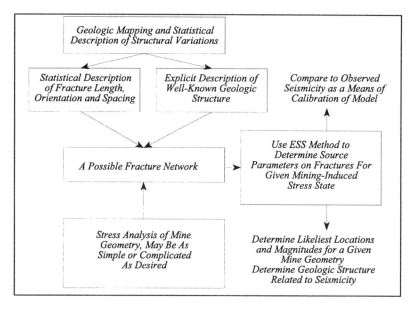

Figure 4. A flow diagram illustrating the primary concepts of the seismic simulation model.

1475

(1) if a fracture slips at some location on its surface, that region is assumed to be in a residual strength condition and not allowed to fail again; and

(2) if a fracture is subjected to tension in the normal direction, non-violent failure is assumed and the fracture is, likewise, not examined further.

The effects of these two restrictions tends to build a "history" into the rock mass as the extraction ratio increases.

The result of this fairly simple procedure is to provide a picture of the potential seismic response of the fracture system without explicitly attempting to model slip on each individual feature. The critical locations and structural orientations for shear failure are determined, and the extent of this failure into the rock mass. The ability of this approach to determine seismic source parameter estimates (e.g., locations, magnitudes, source radii, stress drop, relative shear displacement, etc.) allows the model in be verified against field data from existing waveform digitizing seismic systems. In this way, the model provides a "linkage" between mining methods or sequences and an estimate of seismic potential.

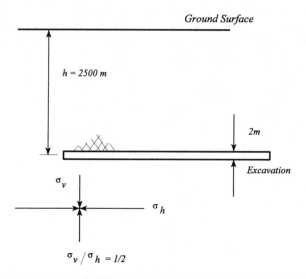

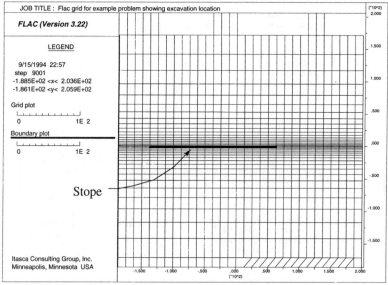

Figure 5. Geometry of example problem and model representation.

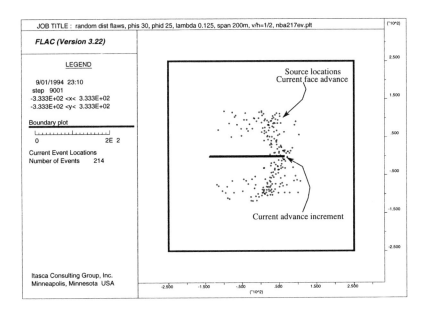

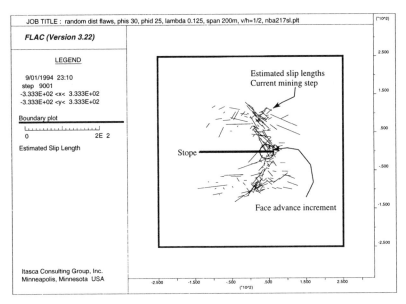

Figure 6. plot of seismic event locations and slipping lengths for a stope span of 150 m.

3.3 Example problem - Seismicity associated with advancement of a deep longwall face

A simple problem is examined in order to demonstrate the approach (Figure 5). A horizontal, thin-reef longwall face at 2500 m depth is advanced in a rock mass subjected to a 2:1 horizontal to vertical stress field. A fracture network characterized by random dip, dip direction and location and by a normal distribution of spacings with a 5 m average is generated in a volume of rock approximately 400 m on a side. The continuous length of the fractures is defined by a negative exponential distribution of fracture radii, with average of 5 m. In this

1477

example, the fractures were all assumed to have the same properties, i.e., cohesionless with a static friction angle of 30° and a dynamic friction angle of 25°. A stope of 1.5 m height was advanced in this rock mass in seven increments from zero to a total stope length of 200 m. At each face advance increment, the stress state was determined from an elastic stress analysis and mapped onto each fracture at a number of locations along the fracture surface, using the technique described above (Figure 6).

Here, the event source locations are shown as well as the slip length of those fractures which have failed. As seen, the source locations arch in a "bow-wave" fashion around the advancing stope face. This is expected, due in the high horizontal stress component. The region directly adjacent to the stope behind the face appears to be free of seismicity; this is the result of plotting only the seismicity resulting from the most recent face advance increment. A common method of represen-ting the seismic response of mining a block of ground is to plot the log of the cumulative number of events with a magnitude exceeding a given value, versus the magnitude (the Gutenburg-Richter plot).

This relationship for the example problem is given in Figure 7. As seen, the plot is roughly linear for large events [defined as those with a magnitude M greater than (approximately) zero], i.e., in agreement with the equation log (N) = a - b M_L, where N is the cumulative number of events, and M_L is the local event magnitude. The intercept, a, is dependent primarily on the volume of excavation, but the

slope, b, is a characteristic value for the rock mass and mining situation. Earthquake seismology records usually indicate b values around 1.0, with mining seismicity reporting values varying from around 0.6 to in excess of more than 1.0. Morrison, et. al., (1993) report a value for b of approximately 0.97 for large events at the Creighton Mine, whereas Spottiswoode and McGarr (1975) determine a "b" value of approximately 0.6 to 0.7 for the ERPM Mine in South Africa. As seen in this example, the slope is approximately 1, corresponding well to field observations. The distributions of stress drop and slip radii (Figure 8) also agree reasonably well with field observations (Gibowicz, 1990). Here, the greatest frequency of stress drops is in the range of 3 to 4 MPa, with values as high as 30 MPa. Field data indicate that values in the range of 0 to 10 MPa are most common. The corresponding slip radii range from less than 1 m for the smallest events to approximately 50 m for the largest events, with relative shear in the range of 1 to 15 mm. Spottiswoode and McGarr (1975) report slip values, estimated from seismological observations, ranging from less than 1 mm to approximately 13 mm at a South African Gold mine.

The effect of stress ratio on the location of seismicity for the same fracture network is given in Figure 9a for the case of a 1:2 horizontal to vertical stress field. As seen here, the seismic source locations are now contained within the regions of increased shearing stress in advance of the face. These locations compared favourably with source

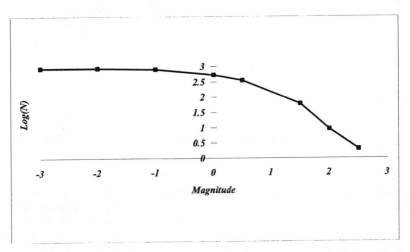

Figure 7. Frequency-magnitude distribution for the base case of ϕ_s = 30°, ϕ_d = 25°, λ = 0.125 (The b-value is approximately 1).

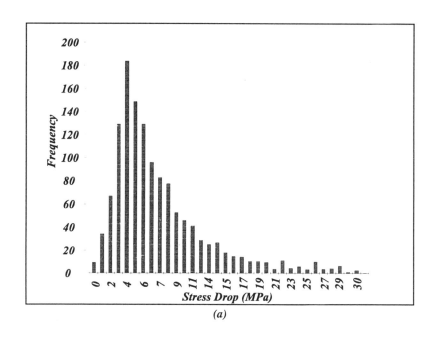

(a)

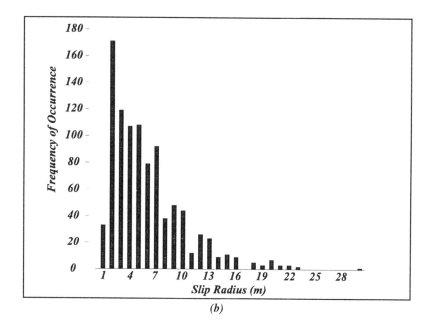

(b)

Figure 8. histograms of (a) predicted stress drop and (b) slip radius for the base case.

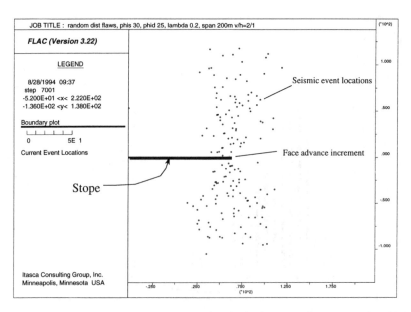

Figure 9a. plot of the immediate face area showing predicted seismic events for a stess ratio of 2:1 vertical-to-horizontal stress.

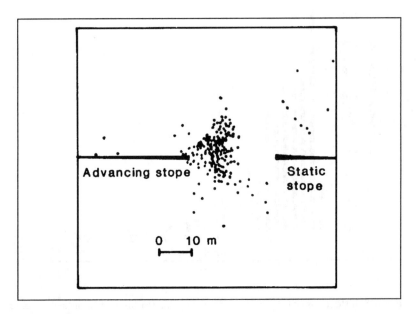

Figure 9b. Seismic event locations in advance of a longwall face for a deep South African gold mine [Legge and Spottiswoode, 1987]

locations observed in South African mines in which a major vertical principal stress is found (see Figure 9b).

A parametric study was completed in which the various fracture geometric and strength properties, in situ stress magnitude, and stress ratio were varied to determine their impact on the estimated seismicity. The factors of greatest influence were found to be the difference between the static and dynamic friction angle of the fracture surfaces (the peak excess shear stress); the continuous length of the fractures and the in situ stress magnitude. The results of these calculations indicate that, using reasonable assumptions of strength and fracture length variation, one can produce a seismic response, which compares well with the observed response. The real advantages of this approach are:

(1) it is verifiable by comparison to field seismological observations at any given mine; and

(2) it attempts to directly tie mine geometry and geology to seismic potential.

The statistical variation in geologic structure can be mapped and described to at least a reasonable level of detail. Assuming the in situ stress state is known, the primary unknown parameters which require calibration are the strength values for the fractures.

4. FIELD EXAMPLE: VERIFICATION OF SEISMIC SOURCE MECHANISMS AT THE LUCKY FRIDAY MINE

4.1 Problem Statement

The Lucky Friday Mine of Hecla Mining Company is located in the Coeur d'Alene mining district of northern Idaho. It is a lead-silver mine which extracts narrow (3 m), near-vertical ore shoots at depths to greater than 1700 m. Currently, the mine uses a longwall undercut-and-fill mining method to extract the approximately 400 m of vein length. Mining progresses in an overall downward direction by breasting beneath the cemented backfill placed in the previous cut. Excavation continues to the next level downward by crosscutting from a number of footwall ramps. The orebody is a relatively hard, siliceous ore, located in the bedded, argillaceous quartzite of the Belt Series of Western Montana and Idaho (Figure 10). The beds range in character from hard, brittle quartzites to softer, impure quartzites and argillites which dip to the southeast at 60°, and are thus intersected by the near-vertical vein. Of particular significance in this problem is a series of

argillitic interbeds (thin, pure argillite beds) which occur every ten meters or so in the bedding sequence. The vein and bedding are conformably folded by a steeply-plunging anticline which gives the orebody a "hooked" appearance. Finally, the vein is terminated on its north and south extremities by two major west-north-westerly striking faults - the North and South Control Faults.

Rockbursting at the mine has been a severe problem since the 1960s, with mining then at depths of less than 1000 m. Until the mid-1980s, overhand cut-and-fill mining was used, creating sill pillars between 65 m levels. Rockbursting would begin when these pillars were reduced to less than about 15 m in height. This prompted the changeover to a longwall system, which eliminated the vertical pillars. In recent years, the mechanics of the rockbursting has been related to unstable slip on the argillite interbeds as well as on the North and South Control Faults (Blake and Cuvelier, 1990). However, to this date, no study has clearly identified the relationship between these structures and the location and frequency of the seismic events.

4.2 Seismic Studies

Since the early 1970s, the Lucky Friday Mine has used a first arrival-only seismic system for locating seismic events in the mine. In the 1990s, this system was supplemented with a digitizing seismic system which has been used to determine source parameters (Lourence, et. al., 1993). Thus, the available seismic data base consists primarily of source locations, and some recent values of magnitude for large (>0 M_L) events. This database indicates that most large seismic events occur in the footwall of the orebody, either in the hook region or along the locations of the terminating fault structures. Figures 11a and 11b show the source locations and magnitudes of events in longitudinal and plan view from a several month period of study in 1992, during which the digitizing seismic system was being tested. In general, events at the Lucky Friday Mine range up to M_L magnitude 4, with magnitude 2 events occurring every one to two months.

4.3 Numerical Analysis of the Seismicity

The seismicity simulation approach discussed above was used to examine the seismicity associated with possible slip on the ubiquitous argillite interbeds as the longwall face was advanced. Slip potential on the terminating fault structures was examined by

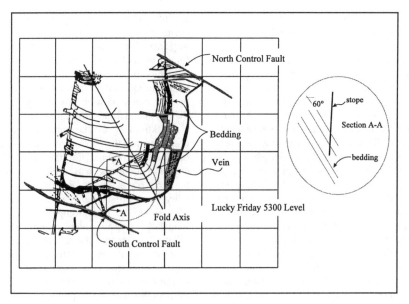

Figure 10. Plan view of the 5300 level (-1615 m) of the Lucky Friday Mine showing the general geology (This map is typical of the geologic mapping performed at the mine.)

explicit modeling of the slip using a three-dimensional discontinuum method (3DEC).

Detailed line mapping of the bedding structures was performed to define their statistical variation in dip, dip direction and spacing; the structures have continuous lengths, varying from 10's to 1000's of meters. As will be shown, in the case of the Lucky Friday Mine, the length of the structures was found not to be the controlling factor in seismic response. A series of two and three-dimensional analyses were conducted in which the longwall advance was simulated over a four year period in one-year increments. The bedding surfaces are generally slickensided and quite planar, and were thus assumed to have zero cohesion.

The seismicity resulting from one face advance increment is shown in the section given in Figure 12. As seen, the seismicity occurs in the western (footwail) of the orebody in a wide arc emanating from the face and extending several hundred meters into the wall. The curious result of the analyses is that the seismicity is restricted primarily to the immediate vicinity of the face or in the western wall. The large seismic events are found almost exclusively in this region, while the minor seismicity in the eastern (hangingwall) is of lower magnitude. This behavior agrees well with the field observation that seismicity is restricted to the footwall region. The modeling implies a source mechanism related to bedding plane slip as illustra-

ted in Figure 13. A shearing mechanism preferentially occurs in the footwall since beds are unfavorably oriented with respect to the stresses (arching) around the longwall face. In the hangingwall side, the beds near the face (where stress drops are large) are clamped by high normal stresses (negative ESS as shown in the previous figure), thereby preventing slip. The slip which does occur in this region of relaxed normal stress behind the face is characterized by fracture opening, resulting in smaller energy releases.

A parametric evaluation was conducted of the frictional properties of the bedding, with static friction values in the range of 25 to 30° and dynamic friction angles of 25 to 28°. The frequency magnitude response is shown in Figure 14. It is seen that the numerically-predicted value of the b slope is approximately 1.1, with a maximum event magnitude restricted to approximately 2.5 to 3 M_L. This restriction in magnitude of the bedding slip events appears to be a function of the stress state and bedding orientation around the face, and not a function of the continuous length of the bedding structures. These conclusions have obvious practical implications, the most important of which is with respect to location of the stope development. Unfortunately, a shaft was sunk in the early 1980s in the footwall to minimize development length. However, the ramp development was placed directly in the region most susceptible to large-magnitude bedding events, thus locating mining in the area of greatest damage potential.

A separate three-dimensional explicit slip analysis was conducted to examine the potential for producing events in excess of magnitude 3 M_L on the terminating fault structures. The 3DEC program (Itasca, 1994) was used for this purpose. The anal-ysis shows that these fault structures are poorly aligned with respect to the northwest orientation of the principal stresses which tend to arch around the ends of the orebody, resulting in large induced shearing stresses on the structures. Modeling was conducted over a span of four years of longwall face advance. When the change-over was made from

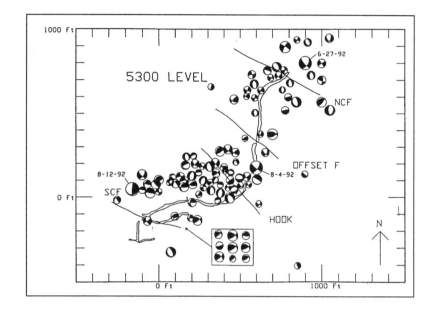

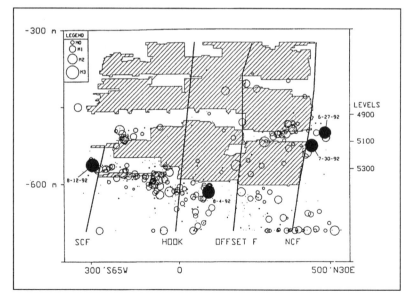

Figure 11. Seismic event locations for the period of May 1 to September 30, 1992, projected onto longitudinal and plan sections the Lucky Friday Mine (the diameter of circle corresponds to magnitude. Major rockbursts are noted; shaded regions are mined and backfilled). [Lourence et al., 1993].

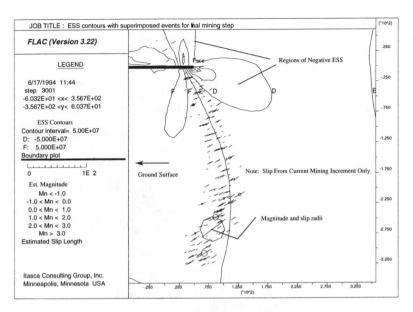

Figure 12. Close-up view of events with superimposed contours of excess shear stress (Events follow the contour of positive ESS).

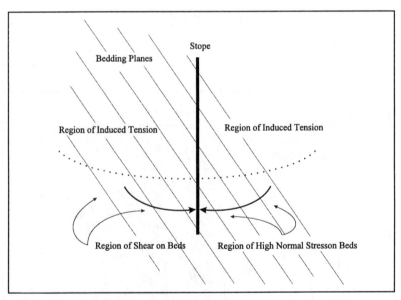

Figure 13. Schematic illustration showing proposed mechanism for seismicity in footwall (northwest wall) (Shearing occurs on bedding planes in left side of orebody while clamping occurs in opposing wall).

overhand to underhand mining, sill pillars were left along the faults. These pillars have proven to be extremely difficult to extract. The modeling showed that these pillars, even though relatively narrow in height, act as local asperities on the faults, and tend to clamp them until they fall violently. The resulting large (>3 M_L) events may be a result of the combined failure of the pillars along the fault traces, allowing violent slip over large surface areas. The practical conclusions to be gained from this analysis are:

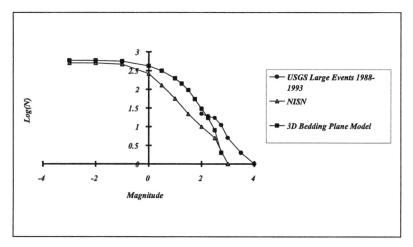

Figure 14. Frequency-magnitude plot for the Lucky Friday Mine, 1992-1993, North Idaho Seismic Network and 3D Model (Large events monitored at the USGS in Newport, Washington, from 1988-1993, are also shown).

(1) that even small pillars left against the faults here can effectively clamp them locally; however,

(2) when they do eventually fall as the face is advanced, excessively large events can result.

It is recommended that fault-locking pillars not be left in place at the Lucky Friday Mine.

CONCLUSIONS

This work was prompted by the need for analytical or numerical methods which could be used to assess the seismic potential of a rock mass when subjected to a specific mining method or sequence. In particular, it was concluded that techniques are needed which:

(1) are based on relatively simple concepts and which can be applied relatively quickly;

(2) help identify the seismic mechanisms and the relative impact of changes to mining methods or geometries on alleviation of the problem - as opposed to attempts to make exact predictions of future rockbursts;

(3) take into account the geology of the mine and the statistical nature of rock mass fracturing; and

(4) can be "calibrated" against data obtained from commonly used wave form digitizing systems.

A numerical methodology for estimation of mine seismicity was discussed here which is based on two fundamental ideas regarding rockbursting, as follows.

1. Rockbursts are primarily the result of shear failure of a rock mass, either along pre-existing discontinuities or within the intact rock itself.

2. Rockbursts are essentially random phenomena, whose occurrence and severity depend largely on the interaction of the mining-induced stress field with the statistical nature of the geometry of rock mass fracturing.

The method developed was shown to produce seismic response which appears to correlate well with observed seismic response. The ESS method, itself, has been criticized as being highly conservative in nature when applied in a general manner. The work presented here illustrates the value in verification of the model against field data in choosing physical properties of the structures to avoid overly-conservative results. Comparison of the model to the observed response of the Lucky Friday Mine indicates that the model can be highly useful in capturing the essence of the seismic source mechanisms, and that it can be used as a basis for making mining decisions.

Finally, the primary objective of this research was to develop an engineering tool which could be used for comparative analyses of seismic potential of various means of mining a block of ground. It is felt that further experimentation with the method must be made before conclusions can be drawn as to the general applicability of this method.

REFERENCES

Aki K., and P.G. Richards. (1980) *Quantitative Seismology, Theory and Methods*. San Francisco: Freeman.

Blake, W., and D.J. Cuvelier. (1990) "*Developing Reinforcement Requirements for Rockburst Conditions at Hecia's Lucky Friday Mine*", in Rockursts and Seismicity in Mines (Proceedings of the 2nd International Symposium, Minneapolis, June 1988), pp. 407-409. C. Fairhurst, Ed. Rotterdam: A.A. Balkema.

Gibowicz, S.J. (1990) "*The Mechanism of Seismic Events Induced By Mining*", in Rockbursts and Seismicity in Mines (Proceedings of the 2nd International Symposium, Minneapolis, June 1988), pp. 3-27. C. Fairhurst, Ed. Rotterdam: A.A. Balkema.

Hedley, D.G.F., S. Bharti, D. West and W. Blake. (1985) "*Fault-Slip Rockbursts at Falconbridge Mine*", in Proceedings of the Fourth Conference on Acoustic Emissions in Geologic Structures (Penn State University, October 1985).

Itasca Consulting Group, Inc. (1994) *3DEC Version 1.5*. Minneapolis: ICG.

Kuiatilake, P.H.S.W., D.N. Wathugala and O. Stephansson. (1990) "*Three-Dimensional Stochastic Joint Geometry Modelling Including a Verification: A Case Study*", in Rock Joints (Proceedings of the International Symposium on Rock Joints, Loen, Norway, June 1990), pp. 67-74. N. Barton and O. Stephansson, Eds. Rotterdam: A.A. Balkema.

Legge, N.B., and S.M. Spottiswoode. (1987) "*Fracturing and Microseismicity Ahead of a Deep Gold Mine Stope in the Pre-Remnant and Remnant Stages of Mining*" in Proceedings of the 6th International Symposium on Rock Mechanics (Montreal, Canada, 1987), pp. 1071-1078. Boston: A.A. Balkema.

Lourence, P.B., S.J. Jung and K.F. Sprenke. (1993) "*Source Mechanisms at the Lucky Friday Mine: Initial Results From the North Idaho Seismic Network*" in Rockbursts and Seismicity in Mines (Proceedings of the 3rd International Symposium, Kingston, Ontario, August 1993), pp. 2 17-222. R.P. Young, Ed. Rotterdam: A.A. Balkema.

McGarr, A. (1981) "*Analysis of Peak Ground Motion in Terms of a Model of Inhomogeneous Faulting*", J. Geophys. Res., 86(B5), 3901-3912.

Morrison, D. M., G. Swan and C. H. Scholz. (1993) "*Chaotic Behavior and Mining-Induced Seismicity*" in Innovative Mine Design for the 21st Century (Proceedings of the International Congress on Mine Design, Kingston, Ontario, Canada,

August 1993), pp. 233-237. W.F. Bawden and J.F. Archibald, Eds. Rotterdam: A.A. Balkema.

Priest, S.D., and J.A. Hudson. (1981) "*Estimation of Discontinuity Spacing and Trace Length Using Scanline Surveys*" Int. J. Rock Mech. Mm. Sci. & Geomech. Abstr., 18, 183-197.

Ryder, J.A. (1987) "*Excess Shear Stress (ESS): An Engineering Criterion for Assessing Unstable Slip and Associated Rockburst Hazards*" in Proceedings of the 6th ISRM Congress, pp. 1211-1214. Rotterdam: A.A. Balkema.

Salamon, M.D.G. (1964) "*Elastic Analysis of Displacements and Stresses Induced by Mining of Seam or Reef Deposits, Part I, Inclined Reef*" J.S. Afr. Inst. Min. Metall., 65, 319-338.

Salamon, M.D.G. (1993) "*Some Applications of Geomechanical Modeling in Rockburst and Related Research*" in Rockbursts and Seismicity in Mines 93 (Proceedings of the International Symposium, Kingston, Ontario, Canada, August 1993), pp. 297-309. R. Paul Young, Ed. Rotterdam: A.A. Balkema.

Scholz, C.H. (1990) "*The Mechanics of Earthquakes and Faulting*". Cambridge: Cambridge University Press.

Spottiswoode, S.M. (1984) "*Source Mechanisms of Mine Tremors at Blyvooruitzicht Gold Mine*" in Rockbursts and Seismicity in Mines (Proceedings of the 1st International Symposium, Johannesburg, September 1982), pp. 29-37. N.C. Gay and E.H. Wainwright, Eds. Johannesburg: S. Afr. Inst. Mm. Metall.

Spottiswoode, S.M., and A.McGarr. (1975) "*Source Parameters of Tremors in a Deep-Level Gold Mine*" Buli. Seis. Soc. America, 65 (1), 93-122.

Starfield, A.M., and P.A. Cundall. (1988) "*Towards a Methodology for Rock Mechanics Modelling*" Int. J. Rock Mech., Mm. Sci. & Geomech. Abstr., 25 (3), 99-106.

Eurock '96, Barla (ed.) © 2000 Balkema, Rotterdam, ISBN 90 5809 843 6

Rock reinforcement systems

C.R.Windsor

Rock Technology, Subiaco, W.A., Australia

ABSTRACT: A terminology for reinforcement practice is proposed based on the idea that a reinforcement device, a rock mass and a reinforced rock mass are all systems of components. It follows that the force-displacement behaviour of each system is a consequence of the particular physical and mechanical characteristics of the system components and their interactions. Using this terminology a method is proposed for the design of reinforcement for excavations in jointed rock. The design method is based on identifying and stabilising all the blocks of rock that can form at the excavation boundary. It uses the early developments in block theory for identification and stability assessment of the blocks. These developments are supplemented with new reinforcement design and assessment procedures for the unstable blocks.

1 INTRODUCTION

Excavations are central to both mining and many civil engineering projects. For economic and safety reasons reinforcement forms an essential component of these projects. This is demonstrated by an estimated world wide usage of in excess of 500,000,000 reinforcement units per annum. Despite significant advances associated with reinforcement hardware, there is still conflict and confusion concerning the advantages and disadvantages of the different devices and their applicability to various rock engineering problems. This is thought to be due to the lack of a formal terminology in reinforcement practice and the lack of rational design procedures for the reinforcement of jointed rock.

The art of rock reinforcement is still developing into a formal engineering discipline and therefore it will take time for the evolution of a formal terminology. However, it is suggested that some of the jargon in current use may be complicating and confusing what are in essence very simple but very important concepts of reinforcement practice. A terminology needs to evolve that includes terms to describe the range of reinforcement hardware, the techniques of applying that hardware to solve rock engineering problems and aspects associated with the installation and performance assessment of reinforcement.

The behaviour of jointed rock is characterised by the nature and disposition of discontinuities. The discontinuities close to the excavation define the surface block assembly and influence its stability. Good support and reinforcement experiences are associated with addressing these instabilities local to the excavation boundaries rather than attempting to modify global instability induced by stresses. A rigorous engineering approach to rock reinforcement design for jointed rock is somewhat intractable because many of the parameters involved cannot be properly quantified. The design methods that take into account a basic rather than a detailed description of the rock mass are particularly popular. Two methods that follow this approach are the precedent design rules (e.g. Lang [1]) and rock mass classification methods (e.g. Barton et al. [2] and Bieniawski [3]). However, experience has shown that reinforcement is most effective under the low stress conditions that accompany surface block instability which leads to the belief that reinforcement schemes for excavations in jointed rock can be designed using a number of simple concepts based on the examination of the assemblage of surface blocks. A methodology is proposed which is built on the foundations of block theory laid down by Warburton [4], Goodman and Shi [5] and Priest [6] which are supplemented with new reinforcement design procedures for unstable blocks.

2 REINFORCEMENT TERMINOLOGY

A terminology is proposed for different aspects of rock reinforcement practice in an attempt to promote a fundamental understanding of the mechanics of rock reinforcement. The proposed terminology attempts to clarify concepts associated with reinforcement behaviour in a manner that is both general and internally consistent. Some of the terminology has been given previously [7] but is repeated here for completeness together with some more recently defined terms [8]. It is hoped that the terminology will help to simplify the design of reinforced rock structures, to formalise the specification of reinforcement installation and to assist in the assessment of reinforcement performance.

2.1 *Rock Improvement*

Rock improvement is a collective term that includes all methods of rock mass treatment that seek to improve the mechanical properties and desirable attributes of a rock mass. Rock improvement includes methods to increase the strength and reduce the deformability of the rock mass, such as rock reinforcement, rock support, ground freezing and pre-consolidation and methods to modify other rock mass attributes such as reduction of pore pressure through drainage and the reduction of permeability through injection grouting.

2.2 *Rock Support and Rock Reinforcement*

The terms support and reinforcement are often used interchangeably. However, it is useful to consider the two terms as being explicitly different due to the method by which they stabilise the rock adjacent to an excavation. Essentially, support is the application of a reactive force at the face of the excavation and includes techniques and devices such as fill, timber, steel or concrete sets, shotcrete, etc. Reinforcement is considered to be improvement of the overall rock mass properties from within the rock mass and will therefore include all techniques and devices installed within the rock mass such as rock bolts, cable bolts and ground anchors.

2.3 *Pre-Reinforcement and Post-Reinforcement*

Pre-reinforcement is the application of reinforcement prior to the creation of the excavation. Post-reinforcement is the application of reinforcement at an appropriate time after the creation of the excavation.

2.4 *Pre-Tensioned and Post-Tensioned Reinforcement*

Pre-tensioning is the application of an initial tension to the reinforcing system during installation. Post-tensioning is the tensioning or re-tensioning of reinforcement systems subsequent to installation.

2.5 *Permanent Reinforcement and Temporary Reinforcement*

The purpose and service life of an excavation dictate the required quality of the reinforcement device and its installation. Permanent reinforcement has an extended service life while temporary reinforcement has a restricted service life.

2.6 *Primary, Secondary and Tertiary Reinforcement*

The preferred interpretation here is associated with the role reinforcement plays in maintaining stability, mainly because the terms cannot be subsequently confused with the concept of permanent and temporary reinforcement. Primary reinforcement is used to maintain overall stability, secondary reinforcement is used for securing medium to large blocks or zones of rock between the primary reinforcing elements and tertiary reinforcement is used in conjunction with surface restraint to prevent surface loosening and degradation.

2.7 *Types of Reinforcement Devices and Techniques*

An extensive review of reinforcement devices and techniques has indicated that three classes of device and technique have evolved over the years:
- Rock Bolts and Rock Bolting (generally less than 3 m in length).
- Cable Bolts and Cable Bolting (generally in the range from 3 m to 15 m).
- Ground Anchors and Ground Anchoring (generally longer than 10 m).

It can be shown that the tensile capacity of commercially available reinforcement is related to the reinforcement element length and this is termed the 'length capacity relationship' of reinforcement.

It is suggested that this relationship has evolved as a 'natural selection' in response to instability problems that find expression over three scales. Furthermore, it may be shown that there is a link between the length - capacity relationship and the notion of primary, secondary and tertiary reinforcement.

3 REINFORCEMENT SYSTEMS

3.1 *The Load Transfer Concept for Reinforcement Systems*

This concept is central to the understanding of reinforcement system behaviour and the action of the different devices and their effects on excavation stability. The load transfer concept can be visualised as being composed of three basic mechanisms:
1. Rock movement and load transfer from an unstable zone to the reinforcing element.
2. Transfer of load from the unstable region to a stable interior region via the element.
3. Transfer of the reinforcing element load to the stable rock mass.
   These mechanisms identify the critical aspects in the design of reinforcement systems.

3.2 *The Reinforcement System Concept*

Rock bolts, cable bolts and ground anchors are all reinforcement systems. A reinforcement system comprises a system of four principal components:
0. The rock.
1. The element.
2. The internal fixture.
3. The external fixture.

The four principal components are shown schematically in Figure 1. Each component of the system is involved in two load transfer interactions. Whilst, the rock is not generally thought of as being a component of the reinforcement system, it has a marked influence on the behaviour of the system and must therefore eventually be considered an integral part of the system. The concept of a 'reinforcement system' is extremely important because the overall behaviour of the reinforcement system is dictated by the aggregated result of the component interactions of the system. Consequently, the concept of a system of principal components holds the keys to understanding the mechanics of reinforcement, the testing for laboratory and in situ mechanical properties of reinforcement and the proper installation of reinforcement.

3.3 *The Fundamental Classes of Reinforcement System*

Within the techniques of rock bolting, cable bolting and ground anchoring there are a large number of rock reinforcement systems available and an equally large number of confusing claims regarding the benefits of each of these, apparently different, systems. Most of the confusion arises from the failure to recognise the basic mechanics of behaviour of reinforcement systems. The load transfer and the reinforcement system concepts may be used to classify the commercial reinforcement devices. This classification results in three fundamental types of reinforcing system:
1. Continuously Mechanically Coupled (CMC) Systems.
2. Continuously Frictionally Coupled (CFC) Systems.
3. Discretely Mechanically or Frictionally Coupled (DMFC) Systems.

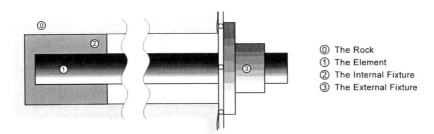

Ⓞ The Rock
① The Element
② The Internal Fixture
③ The External Fixture

Figure 1. The principal components of a reinforcing system.

This terminology has been found to be valid for all reinforcement devices despite the fact that they are available in so many different styles and configurations.

3.4 Reinforcement System Response

The mechanical behaviour of a reinforcement system in response to an excitation or disturbance in the rock mass is termed the Reinforcement System Response. The nature of the response is dependent on, amongst other things, the force-displacement field set up by the disturbance and the arrangement of the reinforcement in that field. For example, the response may be predominantly axial, shear, torsional, or flexural in nature. The most common response is a combination of these modes.

The behaviour of a reinforcement system is determined by the aggregated behaviour of the principal components of the system and their multiple interactions. For example, consider the behaviour of four different reinforcement systems acting within a force-displacement field that results in a simple axial response of the reinforcement. The four different reinforcement systems are formed by various arrangements of four different types of element, four different internal fixtures and four different external fixtures. That is, reinforcement system type X is formed from an arrangement of

reinforcement element type X, internal fixture type X and external fixture type X. The components of the four systems are all subjected to an identical force-displacement field characterised by the force F_S and the displacement d_s. The force-displacement response of the different elements, internal and external fixtures are shown in Figures 2a, 2b and 2c respectively. The responses of the four reinforcement systems are given in Figure 2d. It is worth noting that the response of one or more principal components or the interaction between two components may dictate the overall response of the system. This information enables the design of the individual components to be optimised.

3.5 Types of Reinforcement System Response

Clearly, an infinite number of different reinforcement system responses are possible. However, a reduced number of characteristic types can be defined which are limited by the mechanical and physical properties of the principal components commercially available, the limited possible arrangements of these components to form a particular type of reinforcement system (e.g. Continuously Mechanically Coupled or Discretely Mechanically Coupled) and the limited number of perturbations produced by common excitation mechanisms.

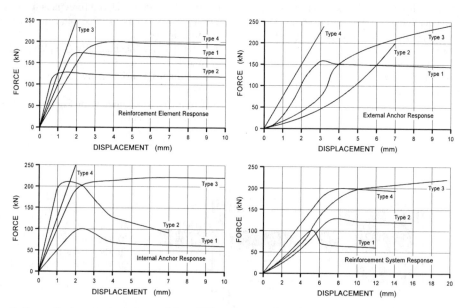

Figure 2. Force-displacement relations of four different types of reinforcing system computed by aggregating the response of the principal components.

3.6 Reinforcement System Capacity

The responses drawn in Figure 2d may be cha-racterised by a number of salient features involving established terminology (e.g. stiffness, yield, ulti-mate, residual and failure). However, two other particularly important terms need to be declared here. System Force Capacity (F_c) is the maximum force that the system is capable of achieving and this has an associated displacement. System Displa-cement Capacity (d_c) is the maximum displacement the system is capable of achieving and this has an associated force.

3.7 Reinforcement System Utilisation and Mobili-sation Factors

The response of a reinforcement system to a given excitation may result in a new state of equilibrium or in failure of the system. If equilibrium is reached, then the force and displacement capability of the system and its individual components are only partly utilised. However, at failure, one or a number of principal components are fully utilised in terms of the force or displacement capability of the system (e.g. element rupture, threads stripped in the external fixture, decoupling of the internal fixture, etc.). Whereas, the force and displacement capacity of other components in the system may have only been partially utilised.

Mobilisation or utilisation can be expressed in terms of the current force or displacement relative to system yield, system capacity or to some other specified design level on the response curve. For example, in terms of capacity, the 'mobilisation factor' indicates the degree to which the force or displacement capacity of the component has been fully utilised. In other words, if the service load and displacement are F_s and d_s then the force and displacement mobilisation factors are given as F_s/f_d and d_s/d_c respectively. Utilisation is equivalent to the mobilisation factor expressed as a percentage.

3.8 Reinforcement Schemes

A Reinforcement Scheme is an arrangement of primary, secondary and tertiary reinforcement systems in a variety of dimensional and spatial con-figurations. Some of these may have been installed as pre- or post-reinforcement and some may be untensioned, pre-tensioned or post-tensioned. The spatial configurations are either uniform geometric arrangements (pattern reinforcement), discretionary arrangements (spot reinforcement) or combinations of the two (combination reinforcement). The performance of pattern and spot reinforcement will be explored later using practical examples. There are two general reinforcement patterns: the rectan-gular array and the oblique array. In each case the array may be dimensioned as 'a x b' where 'a' is the spacing between columns or vertical rings and 'b' is the spacing between rows or horizontal rings. The square and diamond patterns are simply special cases of a = b for the rectangular and oblique arrays respectively.

3.9 Reinforcement Scheme Response

The concepts of system response and system capacity can be equally applied to a 'scheme' because a scheme is really just a global system comprising multiple sub-systems. The behaviour of a reinforcement scheme depends on the aggregated results of the behaviour of natural components of the rock mass (e.g. rock blocks, discontinuities, etc.) and the artificial components (e.g. reinforcement systems, support systems, etc.) and their interac-tions in response to an excitation in the force-displacement field.

Unfortunately, determining reinforcement sche-me behaviour is much more complicated than determining reinforcement system behaviour. In some cases it is possible to estimate reinforcement scheme response by applying analytical and numeri-cal procedures to simple rock mass collapse mecha-nisms (e.g. expansion of a continuum, simple flexu-re of stratified beams, the agitation of simple systems of polyhedral blocks, etc.). However, in many cases formal solution is rendered intractable by the numerous complexities of rock mechanics discussed by Bray [9] which include the severe restriction that proper definition of the problem at hand is often impossible. Consequently, reinfor-cement scheme selection is best attempted by desi-gning on the basis of a simple mechanism. Similar-ly, the response of the scheme is best estimated by analytical or numerical simulation of that mechani-sm. In fact, any predicted scheme response and ca-pacity are intrinsically linked to an assumed mechanism operating within an assumed excitation field. Thus, before discussing the complex mecha-nics associated with reinforcement scheme respon-se, it is first necessary to explore some simple concepts associated with the interactions of the natural and artificial components that make up reinforced rock systems.

4 REINFORCED ROCK SYSTEMS

4.1 *Rock Mass Mechanism Characteristics*

The behaviour of many simple rock mass mechanisms can be described within a 'characteristic force - displacement' field. For example, consider the characteristic force-displacement field for three simple cases:

1. A long, circular tunnel in homogeneous rock under hydrostatic stress - the characteristic force-displacement field is oriented radially and is characterised by (σ_r, δ_r).
2. A slope undergoing circular failure - the characteristic force-displacement field is oriented around the centre of rotation and is characterised by $(M_\theta, \delta_\theta)$.
3. A polyhedral block undergoing translation (i.e. free-falling, sliding on a plane or sliding on a multiplicity of planes) - the field is oriented parallel to the direction of translation and is characterised by (F_t, δ_t).

The force-displacement relationship of a mechanism may be explored using the Mechanism Characteristic Diagram. This diagram displays the various characteristics and relations associated with a particular mechanism in force-displacement space. Consider the simple mechanism shown in Figure 3 which depicts a block with slightly non-parallel faces in the wall of a tunnel. At this stage it is most helpful to ignore the reinforcement and restrict discussion to a mechanism involving the block and the joint surfaces. The block is removable and may slide due to gravity or be driven into the tunnel due to high circumferential stress. In both cases the characteristic force-displacement field is oriented parallel to block translation. A simplified Mechanism Characteristic Diagram for the block is given in Figure 4 and the various characteristics shown on the diagram are now defined.

The Excitation Characteristic is a relation that describes the excitation force-displacement field driving the rock mass mechanism. For the case of a block on a plane under the action of gravity plus an additional driving force (E) the excitation characteristic is displacement dependent until the block detaches from the mass upon which it reduces to a displacement independent, constant characteristic. This constant field is the excitation for the simple gravitational case given by the component of force in the direction of translation. It is thought that most excitation characteristics will comprise some initial, usually non-linear, energy/displacement dependent function that eventually decays to some displacement independent gravitational characteristic.

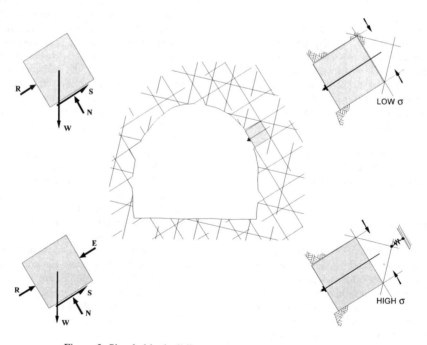

Figure 3. Simple block sliding mechanism in the wall of a tunnel.

The Rock System Response describes the combined force-displacement response of the system of natural components comprising the rock mass. For the case of a block on a plane the rock system response is given by the response of the shearing surface. Again, as with reinforcement system response, the shear surface response is the result of the response of individual components of that sub-system (e.g. shearing through areas of intact rock, areas of filled material, friction on flat surface areas and free translation of unmated surface areas) and their interactions. Consequently, a number of different types of responses are possible. The particular rock response chosen here shows a gradual increase to peak followed by a reduction to a residual strength.

The Mechanism Characteristic is a relation that describes the force-displacement path of the mechanism. This characteristic is determined by considering the interaction of the excitation characteristic and the response of the system of all components acting in the mechanism. In the chosen example, the rock mass system response is insufficient to arrest the mechanism and the Mechanism Characteristic follows the path given in Figure 4. This indicates that an additional contribution from an artificial system of components (e.g. reinforcement and/or support systems) is required to interact with the mechanism to achieve equilibrium. Consequently, the Mechanism Characteristic also indicates the force-displacement requirement of the artificial system - known here as the Artificial System Demand.

4.2 Reinforced Rock System Response

The response of an artificial system of structural components could be drawn on the Mechanism Characteristic Diagram. For example, a small catalogue of reinforcement system responses (i.e. a CFC, a DMFC, a CMC, a tensioned DMFC, and a tensioned CMC (i.e. a tensioned DMFC that has been post-grouted) are shown in Figure 5. Furthermore, the combined artificial/natural response resulting from the interactive effect of including an additional structural component within the mechanism could also be drawn. For example, consider the block shown previously in Figure 3 but now with a reinforcement response included. The Rock System Response, the Reinforcement System Response and the Combined Artificial-Natural System Response (or simply the Reinforced Rock System Response) are shown together with the Excitation Characteristic in Figure 6. The Rock-Reinforcement System Response intersects the Excitation Characteristic at a given force-displacement level and defines the role played by each component in arresting the mechanism and the importance of the relative stiffness of the natural and artificial systems.

4.3 Artificial System Demand

The Mechanism Characteristic Diagram is a useful tool when attempting to decide what type and dimension of artificial system (e.g. shotcrete or

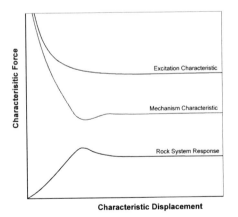

Figure 4. Mechanism Characteristic Diagram for simple block sliding mechanism.

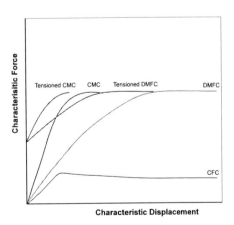

Figure 5. Catalogue of typical load-displacement responses for different reinforcement systems.

1493

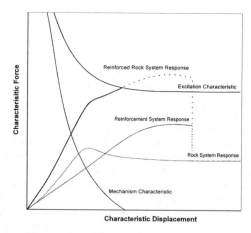

Figure 6. Mechanism Characteristic Diagram for a reinforced block sliding mechanism.

reinforcement, etc.) should be used in response to the demand of a given mechanism (e.g. rock burst, block sliding, swelling etc). Artificial System Demand was introduced earlier as the additional mechanical requirement needed of an artificial system to arrest a mechanism. Clearly, force demand can be determined for any given displacement along the mechanism path. In fact, the force demand used for design may be dictated by a particular design serviceability (displacement) requirement or the need to achieve force equilibrium and displacement compatibility between different 'rock improvement' systems. In other words, a particular reinforcement system demand may result from the requirement to arrest the mechanism at a given displacement. Such constraints dictate the required reinforcement system response and initial conditions.

4.4 Rock Reinforcement Initial Conditions

There are a considerable number of possible variations regarding the timing of reinforcement installation and the stiffness of the reinforcement in relation to development of the rock mass mechanism. In broad terms, the reinforcement may be placed prior to rock mass deformation (i.e. loosening) or at some point during the development of the mechanism. Five main variations on reinforcement timing and initial conditions may be defined; unreinforced, untensioned pre-reinforcement, tensioned pre-reinforcement, untensioned post-reinforcement and tensioned post-reinforcement. The

appropriate choice of reinforcement type (i.e. CMC, CFC or DMFC), its physical and mechanical properties and the choice of initial condition will depend on the rock mass characteristics. For example, the excitation characteristics and the reinforcement requirements for rock burst mechanisms are different to those associated with block type sliding mechanisms.

The effect of system capacity, initial conditions and stiffness for the cases of a pre-reinforced and a post-reinforced simple block sliding mechanism are shown in Figures 7 and 8, respectively. Figure 7 shows the response for a rough joint combined with a number of the reinforcement devices selected from the catalogue of responses given previously in Figure 5. Figure 8 shows the response of a rough joint combined with a reinforcement system (DMFC) placed in various conditions (untensioned (U), lightly tensioned (LT) and highly tensioned (HT)) at different displacements during mechanism development.

4.5 Some Examples of Reinforced Rock System Response

The particular examples discussed to date involve a mechanism where the effect of the response of one component on another is negligible. However, Figure 9 shows that different reinforcement response modes could be generated (e.g. shear, bending, etc), which might modify the response of other components in the reinforced rock system.

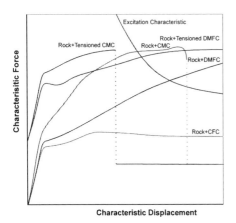

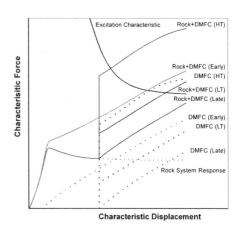

Figure 7. Mechanism Characteristic Diagram for sliding block mechanism post-reinforced with different reinforcing systems.

Figure 8. Mechanism Characteristic Diagram for sliding block mechanism pre-reinforced with different reinforcing systems at different stages of mechanism development.

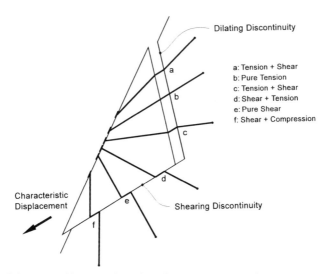

Figure 9. A rock block intersected by a number of reinforcing elements oriented to reinforce different releases surfaces in the rock mass.

Two examples will be explored involving the reinforcement of a cubic block resting on a joint plane inclined at 45°. The first example involves three parallel, but different, reinforcements that interact with the joint but do not affect the normal stress. The second example involves a radial array of similar devices that modify the normal stress

conditions and thus the natural response of the rock component. The response of these two rock reinforcement systems are explored using realistic joint and reinforcement responses and an assumed excitation characteristic. The unit weight of the rock is set to 30 kN/m^3 and the shear strength of the joint is characterised by a peak friction angle of 35°.

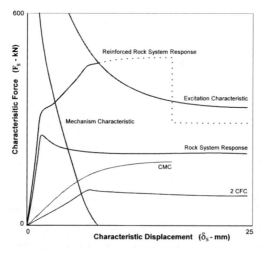
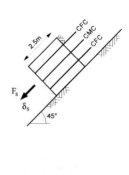

Figure 10. Example of sliding block mechanism reinforced with two different types of rock bolts installed parallel to the inclined joint.

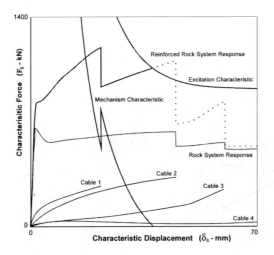
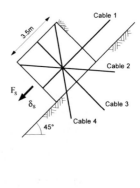

Figure 11. Example of sliding block mechanism reinforced with cable bolts installed at various orientations relative to the inclined joint.

The mechanism characteristic diagram for the first example is given in Figure 10. This case involves a 2.5 metre cube reinforced with two friction stabilisers (CFC devices) and one grouted rock bolt (CMC device) with overall lengths of 3.6 metres all installed parallel to the joint. The Reinforced Rock System Response intersects the Excitation Characteristic at about 8 mm. At this stage, both the joint and friction stabilisers are at residual force capacity while the grouted rock bolt is only utilised to about 80% of its force capacity and 50% of its displacement capacity. Although the fraction stabilisers have only utilised a small fraction of their displacement capacity, it is clear that they alone could never stabilise the mechanism.

The mechanism characteristic for the second example is given in Figure 11. This case involves an array of four, 6 metre long cable bolts that do modify the normal stress conditions on the joint and, thus, the natural response of the rock. In this example it is worthwhile noting the stepped nature of the Rock System Response which reflects the additional normal confinement provided by Cable 2 and Cable 3. Also note that Cable 4 is very ineffi-

1496

cient compared with the other three cables. The Reinforced Rock System Response intersects the Excitation Characteristic at approximately 38 mm. Prior to this state of equilibrium being reached, the cable bolt installed parallel to the plane has failed in tension. This is reflected by the spike shown in the Mechanism Characteristic at about 22 mm.

4.6 *The Mechanism Characteristic Diagram and Reinforced Rock System Response*

In summary, the Mechanism Characteristic Diagram encapsulates the concept of a 'ground response curve' for tunnels proposed by Daemen [10] (excellently described by Hoek and Brown, [11]) and the 'block reaction curve' drawn by Boyle et al. [12]. However, there are a number important differences. In particular, the use of a separate Excitation Characteristic, the use of component system responses within the mechanism and finally the Mechanism Characteristic. Collectively, the Mechanism Characteristic Diagrams indicate a number of important concepts associated with the selection and dimensioning of reinforcement. For example:
a) The natural components of rock mass response may arrest a mechanism.
b) The reinforcement system may be too stiff or too soft depending on the excitation characteristic, the mobilised natural response, reinforcement type and initial conditions.
c) The force 'locked' into the mechanism depends on the excitation characteristic and the equilibrium displacement.
d) The degree to which a sub-component in the rock reinforcement system is utilised is dependent on its component response and the equilibrium displacement.
e) The response of a reinforced rock system may include complete mobilisation or indeed failure of some components.

The Mechanism Characteristic Diagram is simply a graphical way of showing how a system of components undergoing a particular mechanism might simultaneously achieve force equilibrium and displacement compatibility. However, for more complicated mechanisms and improvement schemes, an equivalent numerical approach is required. This is discussed further in Section 7.1.

5 ROCK MASS MECHANISM DEMAND

In order to select and dimension a suitable reinforcement scheme it is first necessary to assume a rock mass collapse mechanism and then determine the artificial demand required to maintain stability of that mechanism. The mechanism assumed will dictate if a rational design process can be implemented and the difficulty in conducting that process. It is probably best to assume a simple collapse mechanism for the rock mass, select a number of candidate reinforcement schemes on the basis of simplified calculations and then conduct a more rigorous analysis on a shortlist of candidate schemes. There are many circumstances in jointed and stratified rock where a simple rock mass mechanism may be assumed and where this approach is appropriate. In order to explain this process in more detail a case in jointed rock involving the instability of blocks of rock will be chosen. Block theory will be used to obtain an initial estimate of the rock mass demand and to explore other concepts associated with the design and response of reinforcement schemes in jointed rock.

5.1 *Force Demand from Block Theory*

In a jointed rock mass the intersections of the discontinuities divide the rock into fully and partially formed blocks of rock. If the discontinuities are assumed to be infinitely continuous then the rock mass is a tightly packed assembly of fully formed blocks. When an excavation plane cuts through this assembly a new set of blocks are formed at the excavation surface. Some of these 'exposed' or 'surface' blocks will have a shape that will allow them to fall, slide or rotate into the excavation should the block driving forces exceed the block stabilising forces. Block analysis is greatly simplified if it is assumed the rock mass is free of stress, the rock material is hard and competent, non-convexities do not occur in the block shape and the blocks behave as rigid bodies that can only translate.

Basically there are two approaches that might be used for defining the block assembly. The main difference concerns the method used to define the spatial positions and occurrences of the discontinuities:
1. If the positions and orientations of each discontinuity and the excavation face are known and defined then a specific block assembly comprising blocks of particular shapes, sizes and positions may be formulated. The vector technique proposed by Warburton [13] is the premier example of the 'specific approach'.
2. If only the general trends in the rock structure are known (i.e. the numbers and orientations of the

discontinuities), then the discontinuities can be assumed to be ubiquitous and a hypothetical block assembly may be formulated comprising all the possible block shapes, sizes and positions. The stereographic technique (plus additional vector analyses) proposed by Goodman and Shi [5] is the premier example of the 'ubiquitous approach'.

The Warburton and the Goodman and Shi methods are equally important in that both cope with arbitrary polyhedra and they are ideal for analysis and design respectively. This discussion is predominantly concerned with design thus a simple ubiquitous procedure will be used to assess the block assembly. The procedure is based on combining the concept of ubiquity given by Cartney [14] and the concept of defining the lists of removable blocks given by Goodman and Shi [5] with the multiplier characterisation for removable blocks given by Delport and Martin [15] and the vector stability analysis technique given by Warburton [13]. It integrates many of the elegant concepts and advances made by these workers with a number of novel procedures introduced to overcome the problems of block size prediction and the definition of the overall design problem. The method incorporates block shape and block size into the stability assessment process. The discontinuities are assumed to be ubiquitous and continuous for the determination of block shape. However, in the determination of block size and block stability both the geometry of the excavation and the dimensional and strength parameters that characterise the rock structure are considered. The procedure has been incorporated within a computer program package called SAFEX which has a modular structure with automatic data file generation and chaining between programs [16],[17].

5.2 Block Shape Analysis

The ubiquitous approach gives rise to a large range of possible block shapes and sizes despite the fact that the analysis is restricted to convex block shapes. For example, consider a rock volume intersected by 'n' continuous, planar and ubiquitous discontinuities and the simple, convex, and removable (i.e. kinematically feasible) blocks of different order (i.e. tetrahedral, pentahedral, etc.) that are formed when a single excavation plane cuts through the rock mass. If r is taken as the number of faces on the block within the rock mass then the total number of removable blocks (TNRB) formed is given by:

$$\text{TNRB} = \sum_{r=3}^{r=n} n!(r-1)(r-2) / \{2\,(n-r)!\,r!\} \qquad (1)$$

If 5 discontinuities are considered then 10 tetrahedral, 15 pentahedral and 6 hexahedral removable block shapes are possible. Furthermore these blocks may occur in a range of sizes from very small to very large.

A computer program called PBLOCK (polyhedral block) has been written to conduct the block shape analysis. In order to define the shapes of the blocks that may form in the faces of an excavation, the orientations of the faces and the discontinuities and whether the rock is situated above or below the face must be specified. The results from PBLOCK comprise a sequence of tables which list the input data, classify the block orders and behaviour modes and give details on each block analysed. For example, consider a flat roof of an underground excavation to be constructed in a rock mass intersected by 5 sets of discontinuities. The tables of results from a block shape analysis conducted by PBLOCK for the roof are given in Figure 12. The last type of table in Figure 12 is the most important in that it lists details on each block shape that are filed for later use and includes:

1. The removable block number.
2. The discontinuities involved in block formation.
3. The block shape code.
4. The mode of block behaviour:
   Type 0 - inherently stable, must be lifted out of the rock,
   Type 1 - inherently unstable, may fall vertically,
   Type 2 - candidate for sliding on a single plane,
   Type 3 - candidate for sliding on two planes,
   Type 4 - candidate for sliding on a multiplicity of planes.
5. Where appropriate, the planes of possible sliding.
6. The direction of block slide (should it prove unstable).

The results tables indicate two important trends. It appears that the general pattern in modes of instability for blocks in each order and the associated translation directions are similar across the different block orders. This trend is also found when excavation faces are rotated through the rock structure [18].

The balance of this discussion concerns tetrahedral blocks because it is suspected that the trends in behaviour of the tetrahedral blocks provides a fairly representative picture of all the block orders. Furthermore, the higher the block order the less likely that the full block will form (i.e. the probability that an n-sided block will form is given by the multiplication of n numbers less than unity associated with each block face). Secondly, the higher the block order, the more chance that the block is stable (i.e. the greater the chance that non-parallel translations on the shear surfaces and minor non-convexities will aid stability). Consequently, the tetrahedral block behaviour may even provide a conservative signature for the rock mass.

5.3 Block Size Analysis

The problem of determining the sizes of blocks that actually form at an excavation face for each block shape is as difficult as the initial problem of determining the shapes of blocks that actually form. In the block size analysis the range of block sizes that could possibly form for each shape will be determined by introducing a 'unit block size' and scaling this between 'limiting block sizes' defined by dimensional data for the excavation and the discontinuities.

The different block sizes will be explained using a typical tetrahedral block shape. The block is shown in Figure 13a annotated to define the four block faces (Face 1 to Face 4), the six block edges

(I_{12} to I_{34}) and the four block vertices (C_{123} to C_{234}). The block face exposed at the excavation surface (Face 4) is shown shaded. Tetrahedral blocks are relatively simple to visualise and define because they have only one vertex completely contained within the rock mass (the apex C_{123}). The position of the apex in relation to the other vertices is fully defined by the orientation of the planes involved in forming the shape. Thus, the size of a tetrahedral block shape may be defined by any characteristic dimension. This allows unit block sizes to be defined with the apex height, face area or volume set to unity. The unit block size may be scaled from infinitely small to four limiting sizes:
1. The excavation limiting block size (face limited block).
2. The trace length limiting block size (trace limited block).
3. The spacing limiting block size (space limited block).
4. The equilibrium limited block size.

The excavation limiting block size is determined by the dimensions of the excavation face. Figure 13b shows the tetrahedral block situated in the roof of an idealised excavation which could be limited by two fixed spans (span 1 and span 2) or be of infi-

POLYHEDRAL BLOCK ANALYSIS	PROJECT: UNDERGROUND	DATE: 08-20-1996
EXCAVATION FACE AT 0 / 0	ROCK MASS SITUATED ABOVE EXCAVATION FACE	

PLANES	PLANE ORIENTATIONS	NORMAL ORIENTATIONS
1	47/135	43/315
2	39/ 10	51/190
3	43/ 87	47/267
4	61/170	29/350
5	28/225	62/ 45

BLOCK TYPES	INFINITE	FINITE	NON-REMOVAB.	REMOVABLE
TETRAHEDRAL BLOCKS	70	82	72	10
PENTAHEDRAL BLOCKS	55	81	66	15
HEXAHEDRAL BLOCKS	16	16	10	6
HEPTAHEDRAL BLOCKS	0	0	0	0
OCTAHEDRAL BLOCKS	0	0	0	0
NONAHEDRAL BLOCKS	0	0	0	0
DECAHEDRAL BLOCKS	0	0	0	0
TOTAL BLOCKS	141	179	148	31

REMOVABLE BLOCKS	BLOCK TRANSLATION MODE					
POLYHEDRAL ORDER	TYPE 0	TYPE 1	TYPE 2	TYPE 3	TYPE 4	TOTAL
TETRAHEDRAL BLOCKS	0	3	5	2	0	10
PENTAHEDRAL BLOCKS	0	3	8	4	0	15
HEXAHEDRAL BLOCKS	0	1	3	2	0	6
HEPTAHEDRAL BLOCKS	0	0	0	0	0	0
OCTAHEDRAL BLOCKS	0	0	0	0	0	0
NONAHEDRAL BLOCKS	0	0	0	0	0	0
DECAHEDRAL BLOCKS	0	0	0	0	0	0
TOTAL BLOCKS	0	7	16	8	0	31

TETRAHEDRAL BLOCK ANALYSIS		10 REMOVABLE BLOCKS			
BLOCK NO.	BLOCK FACE PLANES (DISCONTINUITY NO.)	BLOCK FACE CODE (A) ABOVE (B) BELOW	BLOCK TYPE	SLIDE PLANES	SLIDE DIRECT'N
1	1 2 3	B B A	2	3	43/ 87
2	1 2 4	A B B	3	1 4	45/114
3	1 2 5	B B B	1	-	90/ 0
4	1 3 4	A B B	3	1 4	45/114
5	1 3 5	A B B	2	1	47/135
6	1 4 5	B A B	2	4	61/170
7	2 3 4	B A B	2	3	43/ 87
8	2 3 5	B B B	1	-	90/ 0
9	2 4 5	B B B	1	-	90/ 0
10	3 4 5	B A B	2	4	61/170

PENTAHEDRAL BLOCK ANALYSIS		15 REMOVABLE BLOCKS			
BLOCK NO.	BLOCK FACE PLANES (DISCONTINUITY NO.)	BLOCK FACE CODE (A) ABOVE (B) BELOW	BLOCK TYPE	SLIDE PLANES	SLIDE DIRECT'N
11	1 2 3 4	B B A B	2	3	43/ 87
12	1 2 4 3	A B B A	3	3 1	42/102
13	1 3 4 2	A B B B	3	1 4	45/114
14	1 2 3 5	B B A B	2	3	43/ 87
15	1 2 5 3	B B B B	1	-	90/ 0
16	1 3 5 2	A B B B	2	1	47/135
17	1 2 4 5	A B B B	3	1 4	45/114
18	1 2 5 4	B B B B	1	-	90/ 0
19	1 4 5 2	B A B B	2	4	61/170
20	1 3 4 5	A B B B	3	1 4	45/114
21	1 3 5 4	A B B A	2	1	47/135
22	1 4 5 3	B A B B	2	4	61/170
23	2 3 4 5	B A B B	2	3	43/ 87
24	2 3 5 4	B B B A	2	4	61/170
25	2 4 5 3	B B B B	1	-	90/ 0

HEXAHEDRAL BLOCK ANALYSIS		6 REMOVABLE BLOCKS			
BLOCK NO.	BLOCK FACE PLANES (DISCONTINUITY NO.)	BLOCK FACE CODE (A) ABOVE (B) BELOW	BLOCK TYPE	SLIDE PLANES	SLIDE DIRECT'N
26	1 2 3 4 5	B B A B B	2	3	43/ 87
27	1 2 4 3 5	A B B A B	3	3 1	42/102
28	1 2 5 3 4	B B B B B	1	-	90/ 0
29	1 3 4 2 5	A B B B B	3	1 4	45/114
30	1 3 5 2 4	A B B B A	2	1	47/135
31	1 4 5 2 3	B A B B B	2	4	61/170

Figure 12. Output tables from a PBLOCK shape analysis for an excavation face.

nite extent in either the span 1 or span 2 direction. The maximum block size that can form in the roof face of the excavation is given by the smaller of the span 1 or span 2 limiting block sizes.

The trace length limiting block size is determined by the areal extent of the discontinuities involved in formation of the block shape. For any block shape no line in a block face may be greater than the maximum continuity specified for the joint set associated with that face. In each face of a tetrahedral block contained within the rock mass there are three block edges. A total of nine block sizes may be defined if each edge is taken in turn and set equal to the upper bound on continuity for the associated joint set. The trace limited block is given by the smallest of these nine candidate sizes.

The spacing limiting block size is determined by considering the spacing values of the discontinuity sets. For example, consider the spacing of discontinuities within set number one (Set 1). Figure 13c

shows the example block and three discontinuities that belong to Set 1; the central discontinuity forms Face 1 of the block and the neighbouring discontinuities are situated to the right and to the left of this face to pass through imaginary points at a perpendicular distance of S1 from Face 1. The block is now scaled such that the vertex directly opposite Face 1 (i.e. C_{234}) lies in the plane of the discontinuity to the right. The distance S1 uniquely defines a block size that contains one discontinuity from Set 1. This block size is the S1 limiting block size. Each face is considered in turn to yield block sizes associated with the spacing values of each set (i.e. S1, S2 and S3). The space limited block is the smallest of these three candidates. In the case of the excavation surface also coinciding with the orientation of a discontinuity in the rock mass (e.g. this arises in stratified rock masses) there would be four candidate sizes to consider. The 'maximum block size' that may form in the excavation face is given

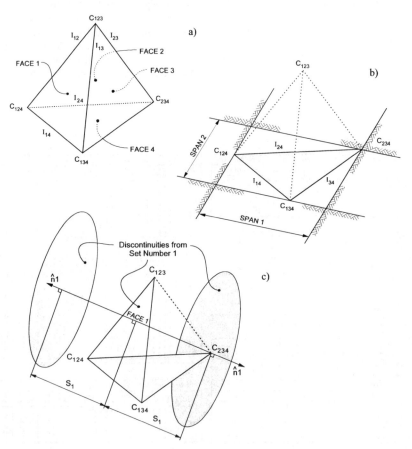

Figure 13. Defining block size using (a) a typical tetrahedral block shape, (b) the excavation span 1 and span 2 limits and (c) the spacing limiting block size.

by either the face limited block size or the trace limited block size, whichever is the smaller. If the space limited block size is smaller than the 'maximum block size' then it represents the 'maximum individual block size' that can form. Under these circumstances the maximum sized block may be made up of smaller, individual blocks. If the space limited block is smaller than the trace limited block size then one pentahedral block and one similar, but smaller tetrahedral block would be formed within the original trace limited block. If the space limited block size is larger than the trace limited block size then this larger shape is only partially formed and the latter remains the maximum size block that can form.

The equilibrium limited block occurs at natural limit equilibrium. That is, the natural disturbing forces acting on the block are just equilibrated by the natural stabilising forces. The equilibrium limited block is not valid for all block shapes (e.g. free failing blocks).

A computer program called SIZE has been written to conduct a size analysis on the list of removable block shapes. When SIZE is initialised it is automatically loaded with a file prepared by PBLOCK. The excavation and discontinuity dimensions may be defined in terms of fixed dimensions or in increments between minimum and maximum values. The operation of SIZE may be demonstrated by returning to the underground example.

The results from the shape analysis have been given previously in Figure 12. The underground excavation is rectangular and measures 40 m x 10 m with the major axis oriented North-South. For convenience, an upper bound on continuity was set to 10 m and a lower bound on spacing was set to 5 m for all discontinuity sets. The results from SIZE are given as a series of tables. For example, the table of results for the first shape in the removable block list (Block No. 1) is given in Figure 14a. This

table sets out the dimensional information for the face limited, trace limited and space limited block sizes. In this case, the trace limited block size with a volume of 15.1 cubic metres is the largest block of this shape that can form in the excavation face and at set spacing values of 5 m it does not contain smaller component blocks. A similar table is given for each block shape in the removable block list and a number of other tables are given that summarise the information for all the blocks. For example, Figure 14b shows a table of volumes for the three limiting sizes for all the removable tetrahedral blocks.

5.4 *Block Shape - Size - Stability Assessment*

With the range of block sizes defined, the stability of each possible block shape over the range of possible sizes may be established. The stability of each block is assessed by conducting a limit equilibrium analysis that takes into account the shape and size of the block, the rock density and cohesive and frictional strength of the translation surfaces. A program module called ANALYSIS has been written to conduct this assessment. ANALYSIS is automatically loaded with standard data files from PBLOCK and SIZE. The user may define the rock density and the cohesion and friction for each discontinuity by choosing the default values, setting specific values or defining a range of values. The stability of each block shape in the list may be assessed in turn. The size of the block may be set to a specific value of apex height, face area or volume or can be set to a unit value of one of these parameters and then scaled between minimum and maximum values.

ANALYSIS was used to conduct a stability assessment on the removable blocks of various sizes that could form in the roof of the underground exca-

PROJECT: UNDERGROUND	SPAN 1 = 40	SPAN 2 = 10	BLOCK No.	1
SET NUMBER	1	2	3	FACE
MEAN SET DIP/DIP DIR.	47 / 135	39 / 10	43 / 87	0 / 0
MEAN SET SPACING	5.000	5.000	5.000	METRES
MEAN SET TRACE LENGTH	10.000	10.000	10.000	METRES
LIMITING BLOCK TYPE	FACE LTD	TRACE LTD	SPACE LTD	
TRACE PLANE 1 IN FACE	13.314	6.090	9.699	METRES
TRACE PLANE 2 IN FACE	10.154	4.645	7.397	METRES
TRACE PLANE 3 IN FACE	11.193	5.120	8.154	METRES
TRACE INT 1/2 IN ROCK	21.861	10.000	15.926	METRES
TRACE INT 1/3 IN ROCK	12.779	5.845	9.309	METRES
TRACE INT 2/3 IN ROCK	15.271	6.985	11.125	METRES
BLOCK APEX DISTANCE	8.550	3.911	6.229	METRES
AREA PLANE 1 IN ROCK	77.825	16.284	41.304	SQ. M.
AREA PLANE 2 IN ROCK	68.981	14.433	36.610	SQ. M.
AREA PLANE 3 IN ROCK	70.163	14.681	37.237	SQ. M.
BLOCK FACE AREA	55.371	11.586	29.387	SQ. M.
BLOCK VOLUME	157.812	15.104	61.016	CU. M.

PROJECT: UNDERGROUND		FACE, TRACE AND SPACE LIMITED BLOCK SIZES				
BLOCK NO.	SPAN 1 (M)	SPAN 2 (M)	CRITICAL SPAN	FACE LTD BLOCK VOL.	TRACE LTD BLOCK VOL.	SPACE LTD BLOCK VOL.
1	40.0	10.0	SPAN2	157.812	15.104	61.016
2	40.0	10.0	SPAN2	9.720	5.072	15.721
3	40.0	10.0	SPAN2	11.637	11.114	6.461
4	40.0	10.0	SPAN2	750.136	1.138	288.738
5	40.0	10.0	SPAN2	153.850	14.755	17.381
6	40.0	10.0	SPAN2	29.901	28.559	48.361
7	40.0	10.0	SPAN2	13.956	6.779	5.678
8	40.0	10.0	SPAN2	21.767	7.696	10.122
9	40.0	10.0	SPAN2	4.156	3.969	18.535
10	40.0	10.0	SPAN2	60.099	21.248	7.222

Figure 14. Results from a SIZE analysis (a) detailed results for Block No. 1 and (b) summary results for Blocks 1 to 10.

vation. The rock density was taken as 27 kN/m³, the friction angle of all discontinuities was set to 25° and the cohesive strength was taken as 0, 2.5, 5.0, 7.5 and 10 kPa for Set 1 to Set 5 respectively. The results table from a stability analysis on Block No. 1 is given in Figure 15a and the results table from a scale-stability analysis on this block is given in Figure 15b.

If the results from the shape, size and stability analyses for all the blocks are combined into single diagrams they effectively characterise the behaviour of the overall rock mass in a way that is compatible with the original input data (i.e. a representative model). One possibility is a scale-stability diagram that summarises the functional relationship between the factor of safety for each block shape and its size. This diagram may be created by plotting a graph of

factor of safety on the vertical axis and one of the characteristics that define block size (e.g. apex height, face area, volume or mass) on the horizontal axis.

For example, a plot of factor of safety versus block free face area for all the tetrahedral blocks for the underground excavation example is given in Figure 16. A horizontal line may be plotted at a factor of safety equal to unity to represent limiting equilibrium. When the planes of sliding are cohesion-less the relations are flat lines at a constant factor of safety; when the block is free-failing the factor of safety is zero. When the discontinuities possess some cohesion the relations reduce with increasing scale and asymptote to a constant factor of safety. Some of these start with factors of safety above unity which decay to below unity. Each rela-

PROJECT: UNDERGROUND	FACE:	0 / 0		BLOCK No.	1
BLOCK APEX DISTANCE	=	1.582	METRES		
BLOCK FACE AREA	=	1.896	SQ. METRES		
BLOCK VOLUME	=	1.000	CU. METRES		
BLOCK MASS	=	2.700	TONNES		
PLANE NUMBERS	1	2	3		
ORIENTATIONS	47/135	39/ 10	43/ 87	DEGREES	
TRACE LENGTHS	2.464	1.879	2.071	METRES	
FACE AREAS	2.665	2.362	2.403	SQ. M.	
PLANE COHESION	0.000	2.500	5.000	KN/SQ.M.	
PLANE FRICTION	25.00	25.00	25.00	DEGREES	
ACTING SHEAR FORCE ON PLANE	3	=	18.064 KN		
ACTING NORMAL FORCE ON PLANE	3	=	19.371 KN		
RESISTING FORCE SHEAR ON PLANE	3	=	21.046 KN		
FACTOR OF SAFETY		=	1.165		

PROJECT: UNDERGROUND	EFFECT OF SCALE ON THE STABILITY OF BLOCK No. 1				
FACE AREA SQ.M.	APEX HEIGHT M.	BLOCK VOLUME CU.M.	BLOCK MASS TONNE	FACTOR OF SAFETY	EQUILIB'M FORCE KN
2.000	1.625	1.083	2.925	1.148	0.000
4.000	2.298	3.064	8.273	0.958	2.330
6.000	2.815	5.629	15.199	0.874	12.823
8.000	3.250	8.667	23.400	0.824	27.584
10.000	3.634	12.112	32.702	0.790	46.028
12.000	3.980	15.922	42.989	0.764	67.762
14.000	4.299	20.064	54.172	0.745	92.496
16.000	4.596	24.513	66.185	0.729	120.008
18.000	4.875	29.250	78.975	0.716	150.117
20.000	5.139	34.258	92.497	0.705	182.673
22.000	5.390	39.523	106.712	0.695	217.551
24.000	5.629	45.033	121.590	0.687	254.643
26.000	5.859	50.778	137.101	0.680	293.854
28.000	6.080	56.748	153.221	0.673	335.100
30.000	6.294	62.936	169.927	0.667	378.309

Figure 15. Stability analysis results on Block No. 1. (a) results for a unit block and (b) results from a scale-stability analysis.

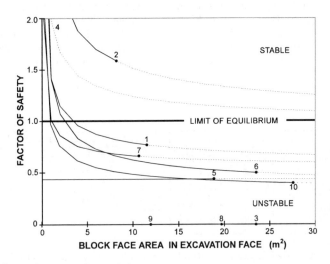

Figure 16. a rock mass scale – stability diagram (block face area versus factor of safety.

tion is truncated by a maximum block size. The area of concern for design is limited to the relations below the equilibrium limit and to the left of the maximum block size limit.

Another characteristic diagram could be drawn that combines 'minimum support pressure' (MSP), 'minimum support length' (MSL) and block face area. MSP is the ratio of the force required to maintain equilibrium of the block to the block free face area. MSL is the apex height of the block plus some allowance for embedment of the reinforcing anchor. A characteristic diagram of this type is shown in Figure 17 for all the removable and unstable tetrahedral blocks that form in the excavation roof. The relations are drawn with a full line and terminated with a spot to represent the maximum size block. The relations for Blocks 3, 7 and 10 are also marked with additional spots which indicate the maximum individual block size limit. Diagrams like Figure 16 and 17 offer the designer the opportunity to select a tentative reinforcement design and this process will be explored in the following section. In practice, the diagrams are usually a little more complicated, especially if a distribution of discontinuities for each set is analysed simultaneously with dispersions for continuity, trace length, cohesion and friction for each discontinuity.

6   SELECTION OF CANDIDATE REINFORCE-MENT SCHEMES

The scale-stability diagram given in Figure 16 could be used to help design the 'rock improvement' system. For example, characteristics often cluster together indicating translations of different shaped blocks with similar displacement vectors. If these groups cross the limit equilibrium line at a very low face area (e.g. less than 0.5 m^2) then mesh or shotcrete might be required to retain the small blocks missed by a widely spaced reinforcement array. If on the other hand no block shapes were found to be unstable at less than a certain face area (e.g. 2 m^2) and if others cross at a greater scale then rock bolts and or cable bolts might be required at a certain array dimension (a x b). Clearly, the size range of unstable blocks can give rise to a variety of surface restraint and reinforcement requirements.

The characteristic diagram given in Figure 17 may be used to estimate a block support pressure and reinforcement length that might be suitable for all block shapes and sizes. Clearly, the length, the force capacity and the array dimensions (a x b) chosen will be more appropriate for some blocks than for others. However, this cannot be assessed until after an initial design has been proposed. A procedure for the selection of candidate schemes is given below and some of the issues associated with scheme assessment are addressed in Section 7.

6.1 *Reinforcement Length, Type. Array Dimension, Capacity and Orientation*

Reinforcement design requires specification of the reinforcement type, length, capacity and reinforcement scheme array dimensions and geometry. A

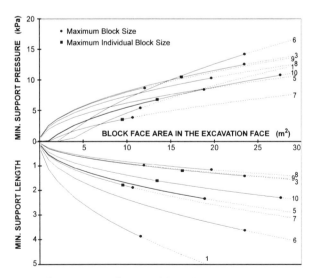

Figure 17. A rock mass scale - support diagram (block face area versus minimum support pressure and minimum support length).

1503

sequence of flow charts describing the reinforcement dimensioning process for block mechanisms has been given previously [19]. One of the most important issues in the selection of candidate reinforcement schemes concerns the commercial availability of suitable products. In fact, the dimension and response spectra of available products form limits on the reinforcement scheme envelope. Other limits may be determined by installation logistics and economic considerations. However, this discussion will be restricted to the selection of candidate reinforcement schemes that fulfil the requirements of mechanism demand.

The minimum required reinforcement length is obtained from Figure 17 (approximately 3.9 m) and in conjunction with the length - capacity relationship indicates that 'rock bolts' are most likely required. Figure 16 indicates that the equilibrium blocks have face areas of about 2 m^2 to 3 m^2. If a 2.0 x 2.0 square array of rock bolts were chosen then mesh would probably be required to retain small blocks (i.e. shapes 3, 5, 8 and 9).

Figure 17 indicates that a minimum block face pressure of about 15 kPa is required. If the required capacity of the rock bolt is taken as the product of

minimum pressure and the array dimensions (i.e. 15 x 2 x 2) then 'reinforcement systems' of at least 60 kN capacity would be required. In a 10 m x 40 m span this results in 20 rings along the north-south axis with 5 reinforcement systems in each ring. The array pressure would be given by (20 x 5 x 60) kN / (10 x 40) m^2 which equals 15 kPa. For DMFC reinforcement systems comprising solid rods, bars and composite tendons a design chart similar to that given in Figure 18 could be used. This particular chart relates the array dimension, array pressure, operating axial stress level in the element to the element area and an equivalent (circular) element diameter. The designer will need to consult the DMFC commercial database and select a possible candidate. For example, one candidate scheme might be a 2 m x 2 m rectangular array of 20 mm nominal diameter Thread bars [7], installed in vertical holes as DMFC devices using resin anchorage internal fixtures, plate, washer and nut external fixtures and torqued on installation to achieve 30% of axial yield (60 kN). The surface restraint data base would indicate that 100 mm x 100 mm light steel mesh (e.g. 6 mm diameter wires) might be suitable for the smaller blocks.

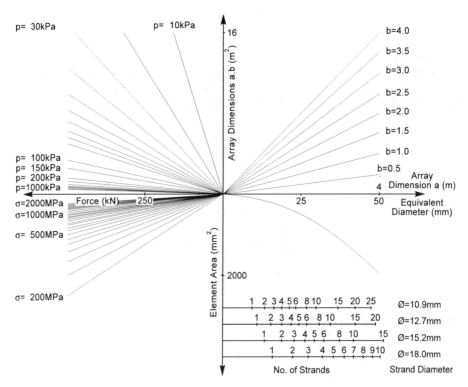

Figure 18. A typical design nomograph for selecting equivalent reinforcement diameter from array dimensions, array pressure and service stress level.

For CMC and CFC reinforcement systems the selection of suitable reinforcement system dimensions are more complicated because the force capacity at the effective point of reinforcement is also dependent on the load transfer achieved in the stable zone above the block. At the tentative design stage it is easier to simply select the appropriate axial capacity device from the CMC or CFC database under the assumption that the required capacity will be available and then assess this later. One candidate scheme could be a 2 m x 2 m rectangular array of 46 mm nominal diameter Split Sets (CFC devices) with deformable dome plates [7] installed in vertical holes. Again, mesh may also be required.

6.2 Initial Geometric and Force Equilibrium Assessment of Candidate Schemes

All the candidate designs need to be initially assessed for each block on the basis of their geometric acceptability and force equilibrium. Geometric acceptability includes checking aspects associated with reinforcement system orientation and length. For example, the orientation and position of the reinforcement in relation to the block shape may mean that the reinforcement does not sufficiently penetrate the stable zone. The required configuration for reinforcement lengths is shown in Figure 19. The total length of reinforcement should be chosen so that the element passes into the stable zone by an amount necessary to achieve a suitable internal fixture length. The internal fixture length to achieve full capacity will vary with reinforcement type. There is a specific minimum requirement for DMFC devices whereas the capacity of a CMC or a CFC device are a function of the internal fixture length.

The block displacement vector and the orientations of the block faces can be used to assess the effectiveness of reinforcement installed at different orientations. The effectiveness of reinforcement may be empirically determined as shown in Figure 20. An effectiveness factor can be used to optimise tension in the reinforcement or simply to rate the efficiency for a particular orientation which may be restricted by other factors such as access or equipment limitations. Consequently, the effective capacity of the reinforcement system may be reduced from its nominal capacity due to reinforcement orientation and the internal fixture length.

Force equilibrium must be checked for each block taking into account the block shape, block size, out of balance force, the approximate number of units to penetrate the block and the effective capacity of the reinforcement system (which may be reduced from the nominal capacity due to reinforcement orientation and internal fixture length). A simple force equilibrium check can be conducted using parameters previously calculated in the block analyses (i.e. shape number (No.), face area (A), apex height (H), volume (V), mass (M), out-of-balance force (OBF) and block face pressure (BFP)). The number of reinforcement penetrations through the block (RN) may be estimated as the integer of A/(a x b) with a minimum value of 1. The

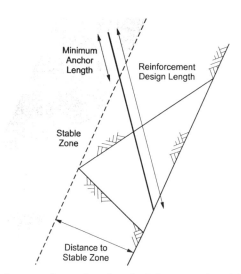

Figure 19. The requirements on minimum anchorage length and reinforcement length in relation to block size.

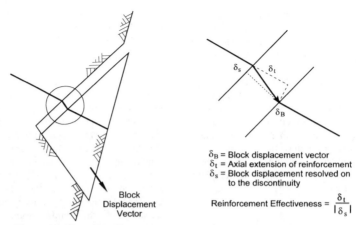

δ_B = Block displacement vector
δ_t = Axial extension of reinforcement
δ_s = Block displacement resolved on
to the discontinuity

Reinforcement Effectiveness = $\dfrac{\delta_t}{|\delta_s|}$

Figure 20. The effectiveness of reinforcement action at a block face.

average embedment length in the stable zone (RL) may be estimated as the overall length of the reinforcement (L) minus half the block apex height.

The reinforcement system capacity (RC) required and other aspects of the required system response can now be calculated for each individual block. For example, the required pre-tension to bring the block to equilibrium could be estimated for DMFC devices. Similarly, the load transfer (RT) required of the internal fixture length could be calculated for CFC or CMC devices. The force-equilibrium calculations and checks on the DMFC devices and the CMC or CFC devices follow similar logic. However, the CFC option will be assessed here because of the more complicated and interesting features associated with the load transfer. The calculations for a force-equilibrium check on

the 46 mm diameter Split Set option are given in Table 1. The blocks that require artificial stabilisation include the free-falling blocks (i.e. shapes 3, 8 and 9) and the sliding blocks (i.e. shapes 1, 5, 6, 7 and 10). This assessment indicates that force equilibrium can be achieved because the usual load transfer capacity of 46 mm Split Sets in correctly sized holes in hard rock is about 30 to 35 kN/m. Furthermore, these calculations indicate that block 6 is of greatest concern and this would be expected from consideration of Figure 17. This block requires high reinforcement system capacity and high load transfer rates to be developed from a minimum of 5 successful installations. Nevertheless this simple force-equilibrium check does indicate that a reasonable good candidate reinforcement scheme has been selected.

TABLE 1. Results from a simple force equilibrium check on blocks in the excavation roof.

No	A (m²)	H (m)	V (m³)	M (t)	OBF (kN)	BFP (kPa)	RN	RC (kN)	RL (m)	RT (kN/m)
1	11.6	3.9	15.1	40.8	63.0	5.4	2	31.5	2.0	15.4
2	8.0	1.9	5.1	13.7	0.0	0.0	2	0.0	3.0	0.0
3	23.5	1.4	11.1	30.0	294.4	12.5	5	58.9	3.3	17.9
4	0.5	6.5	1.1	3.1	0.0	0.0	1	0.0	0.0	0.0
5	18.9	2.3	14.8	39.8	161.5	8.6	4	40.4	2.8	14.3
6	23.5	3.6	28.6	77.1	334.2	14.2	5	66.8	2.2	30.7
7	10.7	1.9	6.8	18.3	42.1	3.9	2	21.0	3.0	6.9
8	19.7	1.2	7.7	20.8	203.8	10.4	4	51.0	3.4	14.9
9	12.0	1.0	4.0	10.7	105.1	8.8	2	52.6	3.5	15.0
10	27.6	2.3	21.2	57.4	298.3	10.8	6	49.7	2.8	17.5

7 ASSESSMENT OF CANDIDATE REINFOR-CEMENT SCHEMES

The procedure used to select and dimension a candidate reinforcement scheme will result in a design that will be more appropriate for some blocks than for others. Unfortunately, there are other issues not yet discussed concerning the force equilibrium - displacement compatibility of each individual reinforced block that may mean this first tentative estimate is inadequate far some blocks. Most of these issues are related to the spatial positioning of the block within the reinforcement array. In fact, many of the geometric issues reduce to a force-equilibrium displacement-compatibility problem. Some of these issues are explored after the requirements of force-equilibrium - displacement compatibility are reviewed.

7.1 Force Equilibrium and Displacement Compatibility

In earlier discussion a reinforced rock mass was considered to be an assembly or system of structural components. It follows that the design of a reinforced rock mass must be considered in terms of force and displacement as the two are inextricably linked in all mechanical systems. Unfortunately, reinforcement is often promoted, selected and designed solely on the basis of force capacity. This is due to, *inter alia*, a number of inappropriate reinforcement design calculations that are cast purely in terms of force such as those given in the previous section. For example, consider the simple equilibrium equation for a structural system of natural and artificial internal and external components:

$$A + B = C \qquad (2)$$

where: A = Artificial Internal and External Reactions (i.e. reinforcement components).

B = Natural Internal and External Reactions (i.e. rock mass components).

C = Internal and External Activating Forces (i.e. perturbations).

The temptation presented by this equation is to assume that all internal and external reactions are the fully mobilised component responses and that the activating forces are a displacement independent excitation. Continuing with these assumptions leads to the Factor of Safety concept, whereby FOS = (A+B)/C. Reference to a Mechanism Characteristic diagram indicates that at limit equilibrium it is extremely unlikely that the component system responses are simultaneously at peak mobilisation.

A useful and more appropriate starting point for determining the reinforcement design might be to assume residual conditions far all natural components, fix the Factor of Safety to unity (the concept is valid at equilibrium) and rearrange the equation to find the required artificial internal reactions to achieve equilibrium as an out of balance force. That is, for when and only when FOS = 1.0:

$$OBF = (FOS * C) - B \qquad (3)$$

This is similar to the approach used previously to check the force-equilibrium for the candidate reinforcement schemes. However, even this approach is generally incorrect because A, B and C are all functions of the rock mass mechanism displacement regime. Accordingly, the equilibrium equation must be rewritten such that Compatibility of Displacement is also properly accounted. For example, this could be written:

$$A (\delta_A) + B (\delta_B) = C (\delta_C) \qquad (4)$$

or simply:

$$F_A + F_B = F_C \qquad (5)$$

For most mechanisms, the solution for the three forces at the equilibrium displacements is most conveniently achieved by rewriting the equation in stiffness-displacement terms:

$$(F_{A0}+K_A\delta_A)+(F_{B0}+K_B\delta_B) = (F_{C0}+K_C\delta_C) \qquad (6)$$

where F_{A0}, F_{B0} and F_{C0} are the initial component forces K_A, K_B and K_C are component stiffnesses and δ_A, δ_B and δ_C are functions of a global displacement δ_g such that:

$$\delta_A = f_A\delta_g$$
$$\delta_B = f_B\delta_g \qquad (7)$$
$$\delta_C = f_C\delta_g$$

The equilibrium equation written in terms of the global displacement:

$$\delta_g = \frac{F_{C0} - F_{A0} - F_{B0}}{K_A f_A + K_B f_B - K_C f_C} \qquad (8)$$

which may be used in an incremental force-displacement procedure to determine F_A, F_B and F_C. This leads to an estimate of the force demanded of the system of artificial components appropriate to the displacement regime of the mechanism, for when and only when δ_g reaches the equilibrium displacement δ_e.

$$F_A = F_C - F_B \qquad (9)$$

If F_A is supplied to the rock mass as a system of artificial reactions then equilibrium is achieved when the displacement of the mechanism reaches δe. Clearly, the preceding discussion is equally applicable to moment-equilibrium and rotation-compatibility.

An incremental force-displacement procedure called WEDGE3D is included in the SAFEX computer program package to assess the stability of an arbitrary reinforced block of any shape. This procedure makes use of the output data from the unreinforced block stability assessment (block shape and block size), material property data (rock density and joint friction and cohesion) and the output data from the design of block reinforcement (i.e. type, orientation, length, capacity and number). To this must be added the load-displacement response of the reinforcing elements obtained from either laboratory or field testing programmes [7]. The block is assumed to be a rigid body, supported by a system of non-deforming boundaries (joint system) and reinforced with a scheme of non-linear axial and shear springs (reinforcement systems). The equilibrium equation is written in terms of force and stiffness and solved for displacement using an incremental Newton-Raphson procedure. The resulting global displacement vector (comprising the three translations and three rotations of the block centroid) are used to back-calculate the component forces at equilibrium and the deformations experienced by each reinforcement system. Example results from some reinforced block stability assessments will be given in the following section.

7.2 Geometric Issues Concerning Reinforcement Length, Orientation and Position

In the assessment process, the analysis must change from a 'ubiquitous' nature to a 'specific' nature. In other words, it is no longer acceptable to assume that the blocks and the reinforcement systems may

occur everywhere and anywhere. The ubiquitous assumption allowed the sum of the forces from all the reinforcement systems to be applied through the block centroid as a resultant force. The assumption that the reinforcement response was independent of displacement allowed this resultant to be set equal to the sum of the completely mobilised force capacities of all the reinforcing systems. It is now necessary to consider the possibility of the various geometric issues and their effect on force equilibrium - displacement compatibility. Three issues will be explored here:

1. The geometric position of the block in relation to the array.
2. The effect of reinforcement length and orientation on load transfer and system response.
3. The effect of block rotation on component responses.

Reinforcement must eventually be defined specifically in position. Figure 21 shows how the number of active elements for the block may vary with position of the face relative to the reinforcement pattern. For the shape given, the number of reinforcing systems sampled is shown to vary from 4 to 2. This problem is addressed here by analysing discrete locations of the block face relative to the reinforcement pattern.

The specific position of the block relative to the reinforcement pattern will also define the effective point of reinforcement action. This point divides the reinforcement length into a block length and an anchor length as shown in Figure 22. The variation in these lengths will affect the response of the reinforcement, especially in the case of CMC and CFC devices.

The previous assumption that blocks can only translate is usually applicable to surface excavations but is not generally applicable to underground excavations. It has also been found that the disposition of reinforcement is less important for sliding blocks formed in surface excavations compared with falling and rotating blocks from overhanging surfaces. In the latter cases, reinforcement will be generally non-uniformly loaded and a simple force equilibrium approach is not valid.

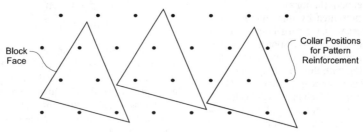

Figure 21. The consequences of specific block position within a reinforcement array.

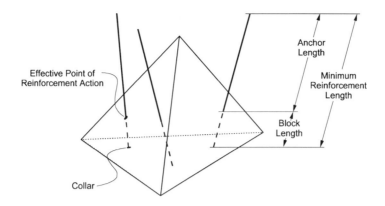

Figure 22. A tetrahedral block and the associated 'specific' reinforcement dimensions.

Reinforcing elements are loaded approximately equal only when the reinforcement is evenly distributed about the block's centre of mass.

These three issues together with a number of other features are explored using three example blocks (blocks 1, 3 and 6). Each block is reinforced with the candidate scheme discussed earlier. The assessment results on block 1, 3 and 6 are given in Figures 23, 24 and 25. The normal and shear reactions from the joints are also reported by the program but have been omitted from these outputs.

Block 1 attempts to slide on discontinuity No. 3 at 43/087. At the trace limited size, block shape 1 can be expected to have about two reinforcement penetrations. However, the block can achieve stability if one reinforcement unit is strategically placed to pass vertically through the block centroid (Case A). Stability is achieved at reasonable force capacity mobilisation and at a translation of 9.1 mm. A reduction in displacement can be achieved by adding an additional reinforcing unit. If the reinforcement elevation is now changed from 90 to 70 degrees, the block becomes unstable (note the reduced embedment length in Case B). Stability can be achieved by moving the reinforcement due east but will be reduced further if moving west. Stability can also be achieved by placing two reinforcement units, both with elevations of 70 degrees through the block (Case C), but again, stability may be lost as they are moved to the west. The important difference between Case A and Case C is the relative magnitude in the shear force developed in the reinforcement at the discontinuity.

Block 3 attempts to free fall (i.e. at 90/000). At the trace limited size, block shape 3 can be expected to have about five reinforcement penetrations. Case A shows that instability occurs even with five reinforcement penetrations with a combined availa-ble axial force capacity of 465.5 kN (taking into account the anchor lengths). On the other hand, Case B shows that stability can be achieved with only 4 reinforcement penetrations given that they are favourable distributed around the centre of mass of the block. However, note that three of the four reinforcement systems have undergone considerable displacement and are close to their available axial force capacities.

Block 6 attempts to slide on discontinuity No.4 at 61/170. At the trace limited size, block shape 4 can be expected to have about five reinforcement penetrations. In both Case A and Case B there are five reinforcement penetrations. The difference between Case A and Case B is that the block is positioned to the west or the east of the array centre respectively. In both cases the combined available axial force capacities are approximately the same (greater than 380 kN). The position of the block means that reinforcement orientations and condi-tions will be slightly different. This leads to instability in Case A and stability in Case B.

The reinforced block assessment process is repeated for each of the blocks in the list of blocks identified as unstable. The reinforcement pattern and system specification must be adjusted and further analyses performed until stability for all blocks is achieved. As expected from the simple check calculations Block 6 does seem to be the block of most concern. However, at this stage it is worth bearing in mind that the design has been conducted at the trace limiting block size. If infinite persistence were assumed for each discontinuity then considerably larger capacity reinforcement would be required to stabilise blocks forming in the roof. For example, at the face limiting volume, block 1 ($157m^3$) is predicted to require 5 cable bolts (1250 kN total axial capacity) to achieve stability.

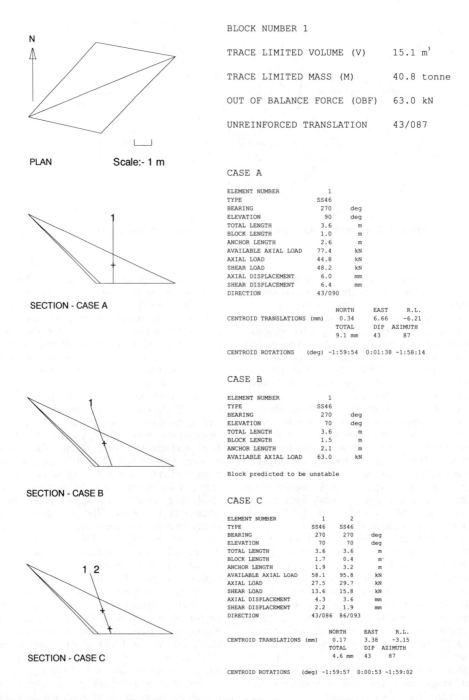

N

PLAN   Scale:- 1 m

SECTION - CASE A

SECTION - CASE B

SECTION - CASE C

BLOCK NUMBER 1

TRACE LIMITED VOLUME (V)   15.1 m³

TRACE LIMITED MASS (M)   40.8 tonne

OUT OF BALANCE FORCE (OBF)   63.0 kN

UNREINFORCED TRANSLATION   43/087

CASE A

ELEMENT NUMBER	1	
TYPE	SS46	
BEARING	270	deg
ELEVATION	90	deg
TOTAL LENGTH	3.6	m
BLOCK LENGTH	1.0	m
ANCHOR LENGTH	2.6	m
AVAILABLE AXIAL LOAD	77.4	kN
AXIAL LOAD	44.8	kN
SHEAR LOAD	48.2	kN
AXIAL DISPLACEMENT	6.0	mm
SHEAR DISPLACEMENT	6.4	mm
DIRECTION	43/090	

	NORTH	EAST	R.L.
CENTROID TRANSLATIONS (mm)	0.34	6.66	-6.21
	TOTAL	DIP	AZIMUTH
	9.1 mm	43	87

CENTROID ROTATIONS   (deg) -1:59:54 0:01:38 -1:58:14

CASE B

ELEMENT NUMBER	1	
TYPE	SS46	
BEARING	270	deg
ELEVATION	70	deg
TOTAL LENGTH	3.6	m
BLOCK LENGTH	1.5	m
ANCHOR LENGTH	2.1	m
AVAILABLE AXIAL LOAD	63.0	kN

Block predicted to be unstable

CASE C

ELEMENT NUMBER	1	2	
TYPE	SS46	SS46	
BEARING	270	270	deg
ELEVATION	70	70	deg
TOTAL LENGTH	3.6	3.6	m
BLOCK LENGTH	1.7	0.4	m
ANCHOR LENGTH	1.9	3.2	m
AVAILABLE AXIAL LOAD	58.1	95.8	kN
AXIAL LOAD	27.5	29.7	kN
SHEAR LOAD	13.6	15.8	kN
AXIAL DISPLACEMENT	4.3	3.6	mm
SHEAR DISPLACEMENT	2.2	1.9	mm
DIRECTION	43/086	86/093	

	NORTH	EAST	R.L.
CENTROID TRANSLATIONS (mm)	0.17	3.38	-3.15
	TOTAL	DIP	AZIMUTH
	4.6 mm	43	87

CENTROID ROTATIONS   (deg) -1:59:57 0:00:53 -1:59:02

Figure 23. The results from stability assessments on Block No. 1.

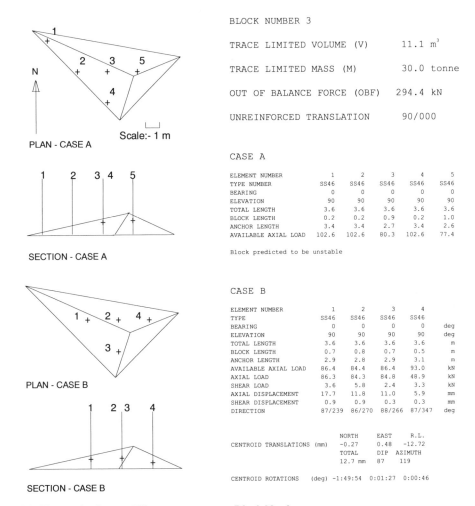

BLOCK NUMBER 3

TRACE LIMITED VOLUME (V)   11.1 m^3

TRACE LIMITED MASS (M)   30.0 tonne

OUT OF BALANCE FORCE (OBF)   294.4 kN

UNREINFORCED TRANSLATION   90/000

CASE A

ELEMENT NUMBER	1	2	3	4	5
TYPE NUMBER	SS46	SS46	SS46	SS46	SS46
BEARING	0	0	0	0	0
ELEVATION	90	90	90	90	90
TOTAL LENGTH	3.6	3.6	3.6	3.6	3.6
BLOCK LENGTH	0.2	0.2	0.9	0.2	1.0
ANCHOR LENGTH	3.4	3.4	2.7	3.4	2.6
AVAILABLE AXIAL LOAD	102.6	102.6	80.3	102.6	77.4

Block predicted to be unstable

CASE B

ELEMENT NUMBER	1	2	3	4	
TYPE	SS46	SS46	SS46	SS46	
BEARING	0	0	0	0	deg
ELEVATION	90	90	90	90	deg
TOTAL LENGTH	3.6	3.6	3.6	3.6	m
BLOCK LENGTH	0.7	0.8	0.7	0.5	m
ANCHOR LENGTH	2.9	2.8	2.9	3.1	m
AVAILABLE AXIAL LOAD	86.4	84.4	86.4	93.0	kN
AXIAL LOAD	86.3	84.3	84.8	48.9	kN
SHEAR LOAD	3.6	5.8	2.4	3.3	kN
AXIAL DISPLACEMENT	17.7	11.8	11.0	5.9	mm
SHEAR DISPLACEMENT	0.9	0.9	0.3	0.3	mm
DIRECTION	87/239	86/270	88/266	87/347	deg

	NORTH	EAST	R.L.
CENTROID TRANSLATIONS (mm)	-0.27	0.48	-12.72
	TOTAL	DIP	AZIMUTH
	12.7 mm	87	119

CENTROID ROTATIONS   (deg) -1:49:54 0:01:27 0:00:46

Figure 24. The results from stability assessments on Block No. 3.

Whereas at the trace limiting block volume (15 m^3) the block can be stabilised with one SS46 rock bolt.

8 CONCLUDING REMARKS

A terminology has been presented for reinforcement based on a systems approach where the reinforcement and the reinforced rock mass are considered to be systems of structural components. The mechanical behaviour of a structural system has been described in terms of component system responses to a given force-displacement field. This approach leads to the Mechanism Characteristic diagram which allows graphical description of the response of reinforced rock systems.

A procedure for the design of reinforced rock

systems has also been proposed for simple instability mechanisms in jointed rock. The component parts of a simple block theory approach and the concepts that link them together have been described and their use has been demonstrated in the design of reinforcement for excavations in jointed rock. The approach combines block shape, block size and block stability analyses in a sequence which has been incorporated in a suite of computer program modules that have been configured to automate a standard sequence of steps requiring minimal user intervention. The modules form part of a design package within a larger package called SAFEX that allows blocks of arbitrary shape, size, position and mode of motion to be analysed with an arbitrary array of reinforcing systems.

Two aspects critical to reinforcement design in

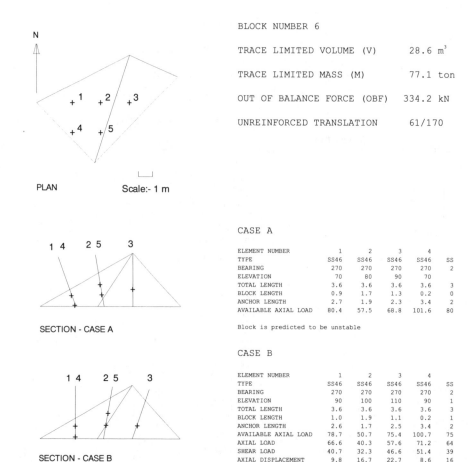

N

PLAN   Scale:- 1 m

SECTION - CASE A

SECTION - CASE B

BLOCK NUMBER 6

TRACE LIMITED VOLUME (V)   28.6 m³

TRACE LIMITED MASS (M)   77.1 ton

OUT OF BALANCE FORCE (OBF)   334.2 kN

UNREINFORCED TRANSLATION   61/170

CASE A

ELEMENT NUMBER	1	2	3	4	
TYPE	SS46	SS46	SS46	SS46	SS
BEARING	270	270	270	270	2
ELEVATION	70	80	90	70	
TOTAL LENGTH	3.6	3.6	3.6	3.6	3
BLOCK LENGTH	0.9	1.7	1.3	0.2	0
ANCHOR LENGTH	2.7	1.9	2.3	3.4	2
AVAILABLE AXIAL LOAD	80.4	57.5	68.8	101.6	80

Block is predicted to be unstable

CASE B

ELEMENT NUMBER	1	2	3	4	
TYPE	SS46	SS46	SS46	SS46	SS
BEARING	270	270	270	270	2
ELEVATION	90	100	110	90	1
TOTAL LENGTH	3.6	3.6	3.6	3.6	3
BLOCK LENGTH	1.0	1.9	1.1	0.2	1
ANCHOR LENGTH	2.6	1.7	2.5	3.4	2
AVAILABLE AXIAL LOAD	78.7	50.7	75.4	100.7	75
AXIAL LOAD	66.6	40.3	57.6	71.2	64
SHEAR LOAD	40.7	32.3	46.6	51.4	39
AXIAL DISPLACEMENT	9.8	16.7	22.7	8.6	16
SHEAR DISPLACEMENT	6.1	13.4	18.4	6.2	9
DIRECTION	58/143	58/143	61/165	54/210	61

	NORTH	EAST	R.L.
CENTROID TRANSLATIONS (mm)	-10.20	3.60	-19.25
	TOTAL	DIP	AZIMUTH
	22.1 mm	61	161

CENTROID ROTATIONS   (deg) 0:12:21 0:02:10 -1:53:02

Figure 25. The results from stability assessments on Block No. 6.

blocky rock have not been covered here:
1. The variable nature of reinforcement system response.
2. The variable nature of rock mass joint system geometry.

In the preceding discussion the reinforcement system response was defined as a force-displacement relation and the reinforcement system capacities as discrete values of force and displacement at particular points on that relation. The response and capacities of reinforcement systems are most commonly obtained by mechanically testing samples of systems or components of systems and then defining a mean response together with associated mean capacities. In reality, there is a statistical variation in the behaviour of a population of similar systems; due in part to slight differences in the individual components and their interactions but more importantly due to variations in the quality of assembly and installation of the system within the rock mass. Basically, the variation in system response may be found by applying to the salient features of the mean system response a sequence of Reinforcement System Modification Factors. The modification factors are themselves functions of a number of parameters and factors. Further explanation is beyond the scope of this discussion but at this stage it is sufficient to recognise that during excavation design the reinforcement system capacities should be modified. For example, to account for the degradation of the system during and after installation, the system force capacity should be modified by a factor which, in most, if not all cases, is less than unity.

In the block shape-size-stability analysis and in the reinforcement scheme proposal and assessment the variable nature of the rock mass was conveniently ignored. The rock structure data was taken as a hypothetical but representative model that defined the geometry of the block assembly and the block procedure reported the results as a hypothetical but representative model that defined the overall range of behaviour of the block assembly. This is thought to be a reasonable starting point that allows an initial set of candidate designs to be proposed. However, a probabilistic approach is required where the elements in the proposed design process are simply 'deterministic engines' within a more appropriate simulation framework. To date, a probabilistic procedure has been completed for convex block shape analysis and current investigations centre on attempting to draw the probability distribution function for block size along the scale-stability relations (giving earlier in Figure 16). The problem of accounting for the block position in relation to the reinforcement array is also under study. The hope is that a probabilistic estimate will be able to be made on the unstable block shape and size ranges and on the likely range in force equilibrium - displacement compatibility conditions. The final answer needs to be expressed in terms of block displacement and reinforcement system forces. Clearly, the analysis will also have to include the probabilistic nature of the reinforcement system response due to quality control issues associated with materials and installation.

ACKNOWLEDGEMENTS

The author wishes to acknowledge all his collaborators in the areas of rock reinforcement and block theory. In particular, he wishes to thank his colleagues Alan Thompson and Glynn Cadby for their constant support, advice and input into all his work. He would also like to thank Mr. G. Worotnicki for providing him with research training, Dr. P.M. Warburton for his guidance and encouragement, Prof. R.E. Goodman and Dr. Gen-hua Shi for introducing him to Block Theory and Prof. S.D. Priest for teaching him the Hemispherical Projection Technique. Finally, but most importantly, the author thanks S.A. Windsor for continually reinforcing and supporting all his endeavours.

REFERENCES

Lang, T. A. (1961). Theory and practice of rock bolting. *Trans. Am. Inst. Mm. Engrs.,* 220: 333-348.

Barton, N.R., Lien, R. and Lunde, J. (1974). Engineering classification of rock masses for the design of tunnel support. *Rock Mech.,* 6: 189-239.

Bieniawski, Z.T. (1976). Rock mass classification in rock engineering. *Exploration for rock engineering,* (Z.T. Bieniawski Ed.), 1: 97-106, Cape Town: A.A. Balkema.

Warburton, P.M. (1981). Vector stability analysis of an arbitrary polyhedral rock block with any number of free faces. *Int. J. Rock. Mech. Mm. Sc. & Geomech. Abstr.,* 18-5: 415-427.

Goodman, R.E. and Gen-hua Shi (1985). *Block theory and its application to rock engineering,* Englewood Cliffs, New Jersey: Prentice-Hall, Inc.

S.D. Priest, (1985). Hemispherical projection *methods in rock mechanics,* London: George Allen & Unwin.

Windsor, C.R. and Thompson, A.G. (1993). Rock reinforcement - technology, testing, design and evaluation. *Comprehensive Rock Engineering - Principles, Practice and Projects* (J.A. Hudson, Ed.) V4, 451-484. Oxford: Pergamon Press.

Windsor, C.R. and Thompson, A.G. (1996). Terminology in reinforcement practice. *Tools and Techniques* (M. Aubertin, F. Hassani and H. Mitri, Eds). Proc. 2nd North American Rock Mechanics Conference NARMS '96. Rotterdam: Balkema, 225-232.

Bray, J.W. (1967). The teaching of rock mechanics to mining engineers. *Transactions of The Mining Engineer,* Institution of Mining Engineers, V126, No. 79, 483-488.

Daemen, J.J.K. (1977). Problems in tunnel support mechanics. *Underground Space,* V1, 163-172.

Hoek, E and Brown, E.T. (1980). *Underground Excavations in Rock.* IMM: London.

Boyle, W. J., Goodman, R.E. and Yow, J.L. (1986). Field Cases using Key Block Theory. *Large Rock Caverns* (K. Saari, Ed.) Proc. Int. Symp. on Large Rock Caverns, V2, 1183-1199: Oxford: Pergamon Press.

Warburton, P.M. (1983). Applications of a new computer model for reconstructing blocky rock geometry - Analysing single block stability and identifying keystones. *5th ISRM Int. Cong. on Rock Mech.,* F225-F230, Melbourne: A.G.S.

Cartney, S.A. (1977) "The ubiquitous joint method, cavern design at Dinorwic power station". Tunnels and Tunnelling, V. 9, No.3, 54-57.

Delport, J.L. and Martin, D.H. (1986). "A multiplier method for identifying keyblocks in excavations through jointed rock". Society for Industrial and Applied Mathematics, J. of Alg. Disc. Meth. Vol.7, No.2, 321-330.

Windsor, C.R. and Thompson, A.G. (1990). *SAFEX user's manual - a design and analysis approach for excavations in rock.* Perth: Rock Mech. Res. Centre, CSIRO Division of Geomechanics.

Windsor, C.R. and Thompson, A.G. (1992). SAFEX - A design and analysis package for rock reinforcement. *Rock Support in Mining and Underground Construction,* (Kaiser and McCreath, Eds). Proc. of Int. Symp. on Rock Support, Sudbury. Rotterdam: Balkema, 17-23.

Windsor, C.R. *(1995).* Block Stability in Jointed Rock Masses. *Fractured and Jointed Rock Masses,* (L.R. Myer, N.G.W. Cook, R.E. Goodman and C.-F. Tsang, Eds.). Proc. of Int. Conf. on Fractured and Jointed Rock Masses. Lake Tahoe. Rotterdam: Balkema, 59-66.

Windsor, C.R. and Thompson, A.G. (1992). *Rock Mechanics* (J.R. Tillerson and W.R. Wawersik, Eds.) Proc. of 32nd US Symposium on Rock Mechanics, Sante Fe. Rotterdam: Balkema, 521-530.

Eurock '96, Barla (ed.)© 2000 Balkema, Rotterdam, ISBN 90 5809 843 6

The foundation of Scott Dam: A case study of a complexly heterogeneous foundation

R.E.Goodman
University of California, Berkeley, Calif, USA

ABSTRACT: The case history of Scott Dam is an engineering tale of trouble that is finally coming to a close with a major repair. This concrete gravity dam was constructed on a weak deformed shale with blocks of hard sandstone that is now known to represent a tectonic melange. Even more difficult geotechnical conditions are to be found immediately upstream and down and on the left abutment. Questions of foundation inadequacy and questionable safety have plagued its owner, Pacific Gas and Electric Company, and with the recent discovery of an active fault only one mile from the dam, it became urgent to reestablish the reserve of safety of the dam under a range of loading conditions. The problem of characterizing the shear strength of melange was attacked by means of fundamental research at the University of California. The results of this research drove a detailed investigation program, which has now led to preliminary design of a strengthening scheme. The lecture will discuss the problems that arose during construction in 1920/21 and in the intervening years, the pre-existing and current model for the shear strength of the foundation, and the recent investigations.

1 SITE CONDITIONS

Pacific Gas & Electric Company's Scott Dam is a concrete gravity dam on the Eel River, in a remote, rugged region of Northern California, near Ukiah. Its reservoir, Lake Pillsbury, provides an important downstream water-supply, as well as nourishing recreation on the Russian River, a fishery, and a small hydro plant.

Despite its modest height, 39.6 m, Scott Dam has been a source of controversy and concern since its construction in 1920/21. The reasons are related to the poor quality of its melange foundation, consi-sting of deformed, fractured shale with blocks of hard graywacke. The melange is contained in a wedge bordered both upstream and down by locally softened serpentinite and on the left side by an active landslide. The foundation is almost entirely underlain by the melange wedge.

2 HISTORY

To build the dam, concrete was poured in monoliths starting from the right abutment and working left. Weathered, softened rock on the right abutment was removed by hydraulic monitor to a depth in places

as much as 5 meters in a somewhat frustrating effort to find rock of more satisfactory condition than softened, deformed shale. A conspicuous outcrop of hard graywacke was the intended foundation abutment on the left bank.

By the end of the 1920 construction season, the dam had not been completed and the rainy season approached. In order to pass the winter storms it was elected to supplement the low level outlet's discharge capacity by channeling the river to flow around the partially completed concrete on its left end; thus flood discharge would run against the left bank. This contradicted the warning of Stanford geology professor J.C. Branner who, as consultant in 1910, noted that the left abutment included conspicuous "landslips ... caused by the removal of the natural supports of the banks"; Branner counseled that this would continue if stream water were allowed to run against the left bank. Unfor-tunately, an unprecedented storm discharge did occur and undercutting of the left bank not only provoked landslide movement, but removed the supposed outcrops of greywacke that were to have formed the right abutment. These outcrops were demonstrated to belong to an isolated block within the melange. The flood transported the block to river level.

To find a possible foundation for the left end of the dam, on resumption of construction in the next Spring, it was elected to rotate the axis of the dam about 450 downstream to pass the dam over a conspicuous pinnacle of hard greenstone. Seams of soft shale and serpentine in the foundation near the pinnacle were removed and replaced with concrete in work completed underground from shafts. The pinnacle was buried in the concrete and construction was completed.

In subsequent years, springs and elevated uplift pressure developed in the downstream, left part of the foundation. Investigations showed that water was flowing from the reservoir in the region of the embedded rock "pinnacle" and a comprehensive drilling program established that in fact this supposed bedrock high constituted instead another isolated rock block. Unfortunately this enormous block had been embedded in the concrete. The possibility of differential movement between the block and the principal foundation, resulting from further landslide movement, was extremely trouble-some to the owners and regulators. Furthermore, shallow earth movements initiated in the slopes downstream of the left abutment. Grouting, draina-ge. construction of a cellular ring wall, and empla-cement of rock fill and filters finally brought the left abutment stability problem to a satisfactory state by 1987.

3 FOUNDATION STRENGTH

Correction of the left abutment problem did not end concern for Scott Dam, however. Seismological instruments revealed a series of small earthquake sources along a trend passing only 1.5 km upstream from the dam. In order to determine the resistance of the dam to strong earthquake shaking, new investigations were inaugurated in 1982 to characte-rize the shear strength of the foundation rock. Four inch (10 cm) diameter samples were taken and tested by Harding Lawson engineers with soil mechanics triaxial testing equipment using both consolidated drained (CD) and consolidated undrai-ned (CU) triaxial tests with pore pressure measure-ment.

Since the foundation rock was highly variable, encompassing large and small blocks in a matrix of softer, shaly rock, it was not a simple matter to assign representative shear strength parameters on the basis of the laboratory tests. Each laboratory test sample yielded markedly different strengths. Ana-lysis of these results suggested grouping the tests into two series, one termed "weaker samples",

consisting predominantly of fractured, highly shea-red shale; and the second termed "stronger sam-ples", consisting predominantly of fractured hard greywacke and siltstone (Volpe et al., 1991) Inde-pendent analysis of the two test subseries yiel-ded effective strength parameters (c', ϕ') equal to 2.0 kPa and 32° for the weaker subseries, and 10.6 kPa and 57° for the stronger series.

In order to combine these two sets of strength parameters, a thorough reexamination of the foun-dation structure was attempted, based on the considerable number of corelogs accumulated over the many years, and new drill holes completed in 1982. A view of the right abutment outcrops from across the valley suggested a rude pattern of folia-tion and a simplified model of the site geology was proposed in which zones of differing proportion of stronger and weaker rocks were recognized. The proportions of stronger versus weaker rock within each such zone were used as weighting coefficients to compute a weighted average shear strength for each trial shear surface passing through the founda-tion.

Although this approach represented a convenient, and workable scheme for assigning shear strengths to different trial slip surfaces, it was appreciated that it could not be supported by geotechnical experien-ce. Therefore, PG&E agreed to enter into a research project at the University of California in order to find the appropriate strength laws for melange type mixtures.

4 A MODEL FOR THE STRENGTH OF ME-LANGE

The author was fortunate to supervise two remar-kable graduate students in pursing this research work: Eric Lindquist, and Ed Medley, Lindquist (1994), Lindquist and Goodman (1994), Medley (1994), Medley and Lindquist (1995), Medley and Goodman (1994). Eric Lindquist identified the fractal nature of the block content of melange as seen in numerous northern California field sites and then created a melange simulant for laboratory triaxial testing. The material consisted of a mixture of Portland cement, fly ash and water for model blocks, and portland cement, bentonite, and water for the matrix material. After initial test results demonstrated that the fundamental variable controlling strength of different mixtures was the volumetric proportion of blocks, different suites of samples were made for comprehensive triaxial testing with differing block proportions. (The attitude of the foliation within the triaxial sample was also varied, but found to be of second order importance as compared with the block proportion.) Lindquist reported a strong increase of effective friction angle with block proportion, equal to approximately 3.0 degrees per ten percent increase in block proportion. He attributed this to the increasing roughness of the failure surface as the number of blocks was increased; the surface of failure invariably remained within the matrix, passing tortuously along the margins of blocks as it averted having to cut through them.

Ed Medley investigated the means of assigning representative values of block proportion based upon results of geological mapping and subsurface exploration. Block proportion could be seen to vary significantly as increasing length of drill core was accumulated, converging to a stationary value, regardless of the order of accumulating the block proportions resulting from different drill holes. Convergence was achieved after a length of drilling equal to about ten times the maximum dimension of the largest block in the volume of rock being investigated. The grain size distribution being fractal, the boundary dimension between "matrix" and "blocks" is dependent on the scale of observation; Medley demonstrated that the boundary between blocks and matrix could be taken, for practical purposes, at a dimension equal to 5% of the square root of the area of observation. He also observed that the maximum block dimension was equal to approximately the square root of the area of observation. (These statistical rules were helpful in planning additional investigations, but require further refinement and testing to know their regions of validity before general adoption.)

5   1995 INVESTIGATIONS

Based upon Medley's statistical observations, it was recommended to acquire an additional total core sample length of about 500 meters from which the block proportion could be evaluated for the rock of the Scott Dam foundation. Based upon Lindquist's experimental observations it was decided to acquire a sufficient suite of high quality core samples to enable a new, comprehensive laboratory testing program to proceed, one which used triaxial compression equipment geared to "weak rock" rather than to "soil". PG&E (1995) conducted these new investigations during the summer dry period in 1995, under the direction of Robert McManus. Laboratory testing was conducted by Dr. Anders Bro of Geotest Unlimited.

Twenty three borings (PQ and HQ sizes) were drilled into bedrock with a total length of 265 meters. The overall percent of core recovery was only 42% despite strenuous attempts to improve recovery. We employed triple tube core barrels, tried drilling with mud instead of water, and experimented with a modified integral sampling technique. We also attempted borehole photography without successful outcome. Various lines of evidence suggested that the material lost from the samples was mainly brecciated blocks, i.e. melange with a relatively high block proportion. Ultimately, a system of determining the block proportion was adopted that combined information from the recovered samples, and from the drilling response in the unrecovered intervals. The block proportion converged to a value of 31% after 150 meters of drilling. (The exploration also tended to reduce confidence in any sort of geologic zonation of the foundation rock, as had been attempted in the previous characterization.)

Sixteen consolidated undrained triaxial tests were conducted on the suite of core samples. Fifteen had block proportions less than 34% and one had a block proportion of 81%. Samples with high block proportions are difficult to obtain because the highly fractured, hard material tends to wash or crumble in sampling. Thus the experimental suite was biased towards the weaker values. Nevertheless, by plotting all results against block proportion, it was possible to interpolate to determine the effective strength parameters corresponding to a block proportion of 31% - the value believed to represent the average condition of the foundation.

The triaxial compression tests yielded strain hardening stress-strain curves. The slope of the plastic portion of the stress-strain curve increased uniformly with increasing block proportion. Because no two samples had the same value of block proportion or the same texture and structure, it was decided to conduct multistage tests; each sample was subjected to three stages of loading at initial effective confining pressures of 275 kPa, 550 kPa, and 1,100 kPa. Pore pressure initially increased in each loading stage, then the rate of pore pressure increase lessened as the sample distorted; in samples with higher block proportion, the pore pressure decrease associated with distortion became greater than the initial pore pressure rise associated with the increasing mean stress, achieving a net negative pore pressure by the end of the loading stage.

6   INTERPRETATION OF CU TESTS

Because of the strain hardening character of the test

results, it was not meaningful to determine the effective strength parameters by fitting an envelope to the three effective stress circles corresponding to any consistent set of points from the three different load stages. Since each test accumulated strain through each of the three subtests, the Mohr's circles for the third loading stage were disproportionately larger than those for the second and first and there was rarely a single envelope even approximately tangent to all three circles. Attempts to fit the three circles yielded some negative cohesion intercepts.

Bro suggested extending the lines in the strain-hardening region of each load stage to achieve pairs of Mohr's circles for a given accumulated strain. Interpreted this way, each test yielded two sets of two Mohr's circles, the envelopes to which had nearly the same friction angle but different effective cohesion values. In other words, the friction angle was constant and independent of strain for a given test while the cohesion increased, approximately linearly, with accumulated strain. The effective strength parameters were thereby deduced, for a value of strain equal to 3%, and plotted against the block proportion. Corresponding to block proportion of 31% the following were the values of the strength parameters: $c' = 290$ kPa; $\phi' = 270$.

A second method used to interpret the results was to construct the effective-stress-path-plot for each sub-stage of the test and collect these to yield a consistent strength curve for each test. This method of interpretation, pursued independently in both p q space and in $\sigma_1' \sigma_3$, space, yielded the following values of the effective strength parameters: $c' = 76$ kPa; $\phi' = 39°$.

The second method gives lower values than the first method for normal stresses less than 713 kPa. Since the average effective vertical stress in the foundation beneath the dam is less than 713 kPa, the second method of interpretation was adopted.

7 CONCLUSIONS

The melange foundation of Scott Dam possesses so much variability that each sample is unique. The distribution of shear strength results was converted to a simple function of one variable by associating a value of volumetric block proportion with each strength value. The overall block proportion appropriate for the dam foundation was determined by a program of drill holes with sufficient length to achieve convergence of the cumulative block proportion.

The values of the shear strength parameters determined by the program of laboratory tests pro-vide an opportunity to evaluate the strength of the dam during an earthquake or an extremely high reservoir condition. The significance and usefulness of these results are apparent when one compares the relative costs of dam strengthening schemes now being considered.

ACKNOWLEDGEMENTS

The author gratefully acknowledges support of Pacific Gas & Electric Company and the contributions of his co-workers within that organization: Charles Ahlgren and Robert McManus. The contributions of Ed Medley, Eric Lindquist, and Anders Bro were referred to in the text. Important earlier work on Scott Dam characterization was conducted by J. Hovland, J. Gambie, and Richard Volpe.

REFERENCES

Volpe, R., Ahlgren, C.S., and Goodman, R.E. (1991). *Selection of engineering properties for geological variable foundations*, Proceedings of the 17th International Congress on Large Dams, Paris, pp. 1087 - 1101.

Lindquist, E.S. (1994). *The strength and deformation properties of melange*, Ph.D. Dissertation, University of California, Berkeley, Department of Civil Engineering.

Lindquist, E.W. and Goodman, R.E. (1994). *The strength and deformation properties of a physical model melange*, in Proceedings of the First North American Rock Mechanics Conference (NARMS), (A.A. Balkema, publishers) pp. 843-850.

Medley, E.W. (1994). *The engineering characterization of melange and similar block-in-matrix rocks*, Ph.D. Dissertation, University of California, Department of Civil Engineering

Medley, E.W. and Lindquist, E.S. (1995). *The engineering significance of the scale-independence of some Franciscan melanges in California, USA*, in Proceedings of the 35th U.S. Rock Mechanics Symposium (A.A. Balkema, publishers), pp. 907-914

Medley, E.W. and Goodman, R.E. (1994). *Estimating the block volumetric proportion of melanges and similar block-in-matrix rocks (bimrocks)*, Proceedings of the 1st North American Rock Mechanics Conference (NARMS), (A.A. Balkema, publishers), pp. 851-858.

Pacific Gas & Electric Company, Geosciences Department (1995). *Geotechnical Investigation*, Scott Dam, Lake County, California, (unpublished; not available).

Keynote lectures

Dr. John Sharp

Professor Roberto Nova

Dr. Vincent Maury

Dr. Bill Dershowitz

Eurock '96, Barla (ed.) © 2000 Balkema, Rotterdam, ISBN 90 5809 843 6

Soft rocks: Behaviour and modelling

Roches tendres comportement et modélisation
Weichgesteine: Verhalten und Modellierung

Roberto Nova & Rocco Lagioia
Department of Structural Engineering, Milan University of Technology (Politecnico), Italy

ABSTRACT: The essential features of the mechanical behaviour of soft rocks are first presented. It is suggested that strain-hardening plasticity, such as the Cam-clay model for soils, is a convenient framework for the description of such a behaviour. In order to get a better match between theoretical predictions and experimental results, a more refined model is presented. Comparisons are shown for a wide range of materials and tests. The possibility of using the model for engineering purposes is shown next.
It is further demonstrated that the theory can be used as a conceptual framework for modelling diagenesis, artificial cementation such as grouting and conversely weathering and weakening due to temperature increase.

RESUME: Les caractéristiques fondamentales du comportement mécanique des roches tendres sont tout d'abord présentées. La théorie de la plasticité avec écrouissage, de type Cam Clay, utilisée pour les sols, est envisagée comme un cadre de reférence convenable, pour décrire leur comportement. Pour améliorer l'accord entre prévisions théoriques et résultats expérimentaux, un modèle plus raffiné est proposé. Plusieurs comparaisons sont présentées pour une large classe de matériaux et d'essais. Les possibilités d'utilisation du modèle pour des problèmes pratiques sont ensuite démontrées.
La théorie peut aussi être utilisée comme cadre conceptuel pour la modélisation de phénomènes comme la diagénèse, la cémentation artificielle dûe aux injections de coulis mais aussi de phénomènes tel que l'altération et la perte de résistance dûe à l'augmentation de la température.

ZUSAMMENFASSUNG: Die wichtigsten Merkmale des mechanischen Verhaltens von Weichgesteinen werden zuerst präsentiert. Es wird vorgeschlagen, dass die Plastizität der Dehnungsverfestigung, wie das Cam-Clay Modell für Boden, ein passendes Gefüge für die Beschreibung eines solchen Verhaltens darstellt. Um eine besseres Anpassung zwischen den theoretischen Voraussagen und den experimentalen Ergebnissen zu erhalten, wird ein verfeinertes Modell präsentiert. Vergleiche für einen umfangreichen Bereich von Materialien und Tests werden aufgezeigt. Die Möglichkeit das Modell für Zwecke des Ingenieurwesens zu verwenden, wird als nächstes demonstriert.
Weiterhin wird aufgezeigt, dass die Theorie als ein Begriffsrahmen für die Diagenese, künstliche Zementierung wie Injektion und zum Gegensatz Verwitterung und Entfestigung aufgrund eines Temperaturanstieges, verwendet werden kann.

1 INTRODUCTION

With the term 'soft rocks' we usually refer to a broad class of geomaterials which lie in the grey area of transition between soils and rocks, as commonly perceived by geotechnical engineers: soils seen as granular media with little or no bonding between particles and rocks considered as hard and strong materials, whose engineering behaviour is governed more by the properties of the discontinuities pervading the rock mass than by the intact rock properties themselves.

This transition zone is often designated also with the term 'weak rocks', Fig. 1, and a clear distinction is made between soils, weak rocks and rocks tout-court on the basis of the value of the (characteristic)

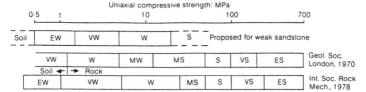

Figure 1. Definitions for weak rock (modified after Dobereiner and De Freitas (1986)).

uniaxial strength of the geomaterial (Dobereiner and De Freitas 1986).

With the term soft rock, however, it is now come into fashion (see e.g. Anagnostopoulos, Schlosser, Kalteziotis and Frank (1993)) to refer to a less clearly defined class of geomaterials, mostly belonging to the set of weak rocks, but comprising bonded soils and, for special problems, even hard rocks weakened by weathering or temperature effects. The common feature of the class of soft rocks is that of being 'intact' materials, in the sense that they can be considered as continua, and that their behaviour is similar to that of hard rocks at low confining pressures and similar to that of soils for higher confining pressures. A rock which is weak because of intense fracturation or existence of cleavage planes may not be a soft rock in the sense above.

To the class of soft rocks belong geomaterials, usually but not exclusively, of sedimentary origin, for instance marls, calcarenites, tuffs, shales, chalks. Limestone, slate or saprolitic rocks can also be considered as soft rocks even though their uniaxial strength is generally higher than that of the aforementioned materials. On the other hand grouted and naturally cemented soils also belong to this class, not withstanding that their uniaxial strength is considerably lower than that of a weak-rock.

2 EXPERIMENTAL EVIDENCE

A typical example of a soft rock behaviour is given in Fig.2 after Aversa, Evangelista and Ramondini (1991). The geomaterial tested in triaxial compression is a fine grained tuff from the Neapolitan area. At low confining pressures the tuff behaves qualitatively as a hard rock: the stress-strain relationship is quasi-linear up to a peak, which is followed by a sudden drop in strength until a final plateau is reached. With increasing confining pressure the behaviour becomes ductile and the

deviatoric stress increases with strain even after linearity (and reversibility) is abandoned.

A qualitatively similar behaviour is showed by a series of specimens of oolitic limestone (Elliott and Brown 1985) Fig.3. In this case the transition from brittle to ductile behaviour is accompanied by a transition from dilative to contractive behaviour, which could be perceived also in the former case, even though in a less clear way.

Both sets of data could be interpreted qualitatively having the Cam-Clay model in mind (Schofield and Wroth 1968). According to this model, the behaviour of an overconsolidated clay can be described as follows: the process of preconsolidation generates a domain in the stress space within which the behaviour of the material is essentially reversible and can be considered as elastic. When the stress path reaches the boundary of this domain, which in the Roscoe and Burland (1968) version of the model is an ellipse, Fig.4, in the triaxial plane p', q, plastic, i.e. irreversible, strain may occur. If the overconsolidation ratio is low, i.e. the isotropic pressure p' is high with respect to the preconsolidation pressure, as the deviatoric strains, ε, increase also the deviatoric stress, q, increases until an asymptotic value is achieved. Volumetric strains, v, are contractive. The Cam-Clay model assumed in fact the validity of the so called normality rule, which implies that the plastic strain rate vector, $(\dot{v}^p, \dot{\varepsilon}^p)$ is directed as the normal to the yield locus, which is the boundary of the elastic domain. It is evident from Fig.4 that for high pressures plastic volumetric strains are positive. The elastic domain increases in size along with plastic volumetric strains.

Conversely, if the overconsolidation ratio is high, so that the isotropic pressure is low with respect to the preconsolidation pressure, when the yield locus is reached, plastic volumetric strain rates are negative: an increase in volume of the specimen is linked to a weakening of the material, the elastic domain shrinks and the deviator stress decreases.

By comparing the model behaviour with the aforementioned experimental results it is apparent

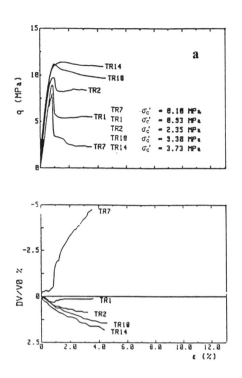

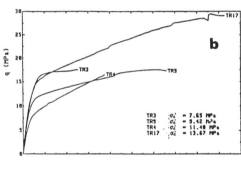

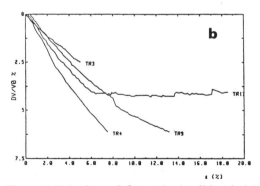

Figure 2. Behaviour of fine grained tuff in triaxial compression: a) low confining pressures b) high confining pressures (after Aversa et al. (1991)).

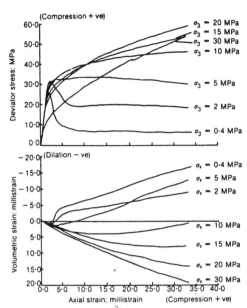

Figure 3. Brittle-ductile transition as a function of confining pressure of oolitic limestone (after Elliott and Brown (1985)).

that the only qualitative difference lies in that for soft rocks, contrary to soils, there is no need of a preconsolidation for the existence of an initial elastic domain. For soft rocks the role of the preconsolidation pressure, p_c in Fig.4, is played by the isotropic pressure at which a marked transition between linear elastic to elastoplastic behaviour occurs under isotropic pressure. Fig.5 shows for instance the behaviour of three types of chalk in hydrostatic tests where p_c varies from 15 MPa to about 40 MPa depending on the initial porosity of the material. Such a pressure can be related to the strength of the bonds between grains and the porosity of the specimen.

From a quantitative stand-point, however, the Cam-Clay model is too simplistic for modelling the complex behaviour of soft rocks. If, for instance, we try to fit the yield points for the tuff, we get a curve like that shown in Fig.6 (Nova 1992) which is quite different from an ellipse, especially for low isotropic pressures. In addition, neither the compactant behaviour for low pressure shown by tuff, nor the partially dilative behaviour occurring for limestone for pressures close to the brittle-ductile transition could be modelled, as well as many other features. In particular, the pore collapse phenomenon was observed for high porosity chalk by Addis and Jones (1989) and for a calcarenite by Lagioia and Nova

1523

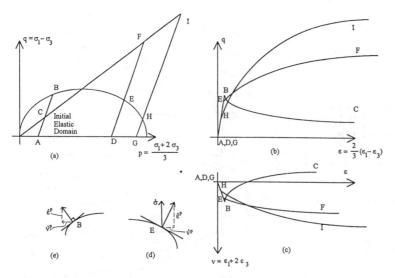

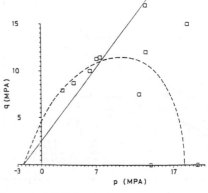

Figure 4. Schematic picture of the Cam Clay model: a) stress-paths in constant cell pressure tests with different overconsolidation ratios, b) stress-strain behaviour, c) volumetric strains, d),e) particulars at points E and B showing the effects of normality rule.

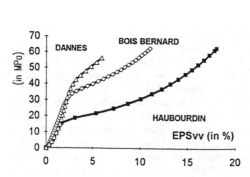

Figure 5. Hydrostatic behaviour of three types of chalk (after Siwak et al. (1993)).

Figure 6. Yield locus for the tuff of figure 2 (after Nova (1992)).

(1995) occurring at constant stress and constant strain rate, Fig.7.

Destructuration has important consequences on the stress path in an oedometric test, as shown in Fig.8 (Lagioia and Nova 1995) and as observed on specimens of artificially bonded soils (Maccarini 1967, Coop and Atkinson 1993), and chalk (Addis and Jones 1989).

Note, however, that destructuration does not necessarily implies degradation of the elastic properties. In the constant cell pressure test shown in Fig.9 on Gravina calcarenite it is shown that the elastic stiffness remains unchanged despite the apparent process of destructuration and the number of unloading-reloading cycles performed. It is also noteworthy that, after destructuration, the yield stress depends on the preloading level, as for virgin soil.

In the following section we shall try to establish the basic structure a mathematical model aiming at describing such experimental evidence should have. A short presentation will be given next of a particular model and various comparisons will be presented between model predictions and experimental data for

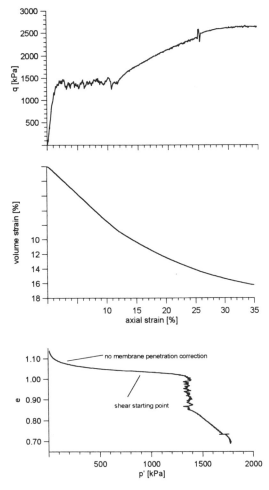

Figure 7. Yielding of Gravina calcarenite in a drained constant cell pressure test (after Lagioia and Nova (1995)).

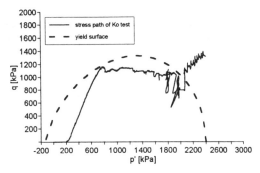

Figure 8. Oedometric stress path on Gravina calcarenite (after Lagioia and Nova (1995)).

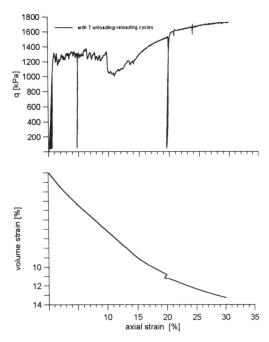

Figure 9. Persistence of elastic stiffness despite degradation (after Lagioia and Nova (1995)).

various types of soft rocks. Examples of the possible uses of such model will be eventually discussed.

3 STRUCTURE OF A BASIC MATHEMATICAL MODEL

The behaviour of soft rocks described in the previous section may look at first glance quite complex: it is non linear, irreversible, path dependent, depends on the geological as well as the mechanical previous history. Moreover it depends on time, strain rate, temperature and orientation of the specimen with respect to the frame of principal stresses. A complete mathematical description is therefore of overwhelming difficulty.

Some aspects are more important than others, however. In the opinion of the writers, the most important is the difference in behaviour before and after a certain threshold is reached.

When a specimen is loaded, no matter what is the stress path followed, it is initially rather stiff, and strains are almost completely recovered in unloading.

However, if a certain threshold (stress path dependent) is reached, then the behaviour becomes non-linear and permanent strains remain after unloading. Some collapsible rocks manifest a sharp transition, since a marked change of void ratio takes place at constant state of stress. For other types of rock the transition is smoother. The change of regime is apparent in both cases, however.

In order to model such two-regimes behaviour we need first a criterion to delineate between one regime and the other. Since we note experimentally that the threshold depends on the state of stress and not on the way such state is achieved, provided no other threshold point is reached, the required criterion will be expressed as a scalar function of the stress state σ_{ij}. If σ_{ij} is within a closed domain in the stress space

$$f\left(\sigma_{ij}, \psi_k\right) < 0 \qquad (1)$$

the behaviour of the rock will be considered to be elastic, Fig.10a.

The vector ψ_k appearing in Eq.(1) is a vector of n hidden variables that should be experimentally determined. In the simplest case $k = 1$, and $\psi_1 = p_{co}$ controls the size of the domain in which the behaviour is elastic.

Clearly, the shape of f itself should be experimentally determined in principle.

If

$$f\left(\sigma_{ij}, \psi_k\right) = 0 \qquad (2)$$

then irreversible behaviour may take place. Since Eq.(2) is the equation of a closed surface in the stress space, it will be called in the following first yield surface (or locus).

The fulfilment of Eq.(2) is not by itself sufficient to cause permanent strains, for non collapsible rocks at least. We may touch the yield locus and unload immediately after. No permanent strain will occur in that case.

We need then more information, that will depend on the direction of the stress rate $\dot{\sigma}_{ij}$. If after the increment the stress state will be within the domain delimited by the yield surface, Fig 10b, it is reasonable to assume that only elastic strains take place. This will occur if:

$$\frac{\partial f}{\partial \sigma_{ij}} \dot{\sigma}_{ij} < 0 \qquad (3)$$

where, conventionally, the repetition of indices implies summation. Note that this is a hypothesis.

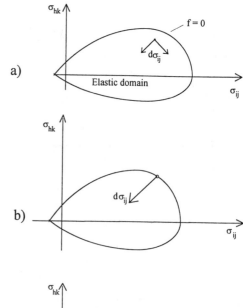

Figure 10. Elastic domain and its evolution: a),b) stress rates giving rise to elastic strains, c) stress rate causing plastic strains and change in size and/or position of yield locus.

Indeed we will be compelled to relax it in the following. Other possibilites exist, but the coupling of conditions (2) and (3) is the simplest way of dealing with unloading.

On the contrary, if

$$\frac{\partial f}{\partial \sigma_{ij}} \dot{\sigma}_{ij} > 0 \qquad (4)$$

elastic as well as plastic, i.e. permanent strains, will take place. We may observe experimentally that when this occurs the yield locus must change size, shape and/or position, Fig.10c. If we unload a specimen after yielding and reload, we observe that new yielding takes place at a point which is outside the initial yield surface. If the reloading path is the same as the unloading, the new yield point is very

1526

close to the point at which unloading started and can be considered coincident with it for all practical purposes.

A stress point on the new yield surface will fulfill again eq.(2), where the parameters ψ_k have changed. In general they will be function of the plastic strains experienced:

$$f\left(\sigma_{ij}, \psi_k\left(\varepsilon_{ij}^P\right)\right) = 0 \tag{5}$$

Since f does not change value when plastic strains occur, the total differential of f should be nil:

$$\dot{f} = \frac{\partial f}{\partial \sigma_{ij}} \dot{\sigma}_{ij} + \frac{\partial f}{\partial \psi_k} \frac{\partial \psi_k}{\partial \varepsilon_{rs}^P} \dot{\varepsilon}_{rs}^P = 0 \tag{6}$$

Note that f can not take positive values since f is initially non-positive and $\dot{f}$ is negative when $f = 0$ and only elastic strains take place, while $\dot{f} = 0$ when $f = 0$ and elastic as well as plastic strains occur.

We have now, formally, a criterion for deciding whether plastic strains will occur or not, but we do not know yet how large such strains will be. In order to determine them, we have to fulfill Eq.(6) and the so called continuity condition, which states that strain rates should vary continuously with stress rates even in passing from elastic to the elastoplastic regime. This can be accomplished by imposing that for neutral loading:

$$\frac{\partial f}{\partial \sigma_{ij}} \dot{\sigma}_{ij} = 0 \tag{7}$$

all the independent components of the plastic strains rates are zero. Note that Eq.(6) could be fulfilled even by a different combination of $\dot{\varepsilon}_{rs}^P$.

In order to do that, it is convenient to assume that $\dot{\varepsilon}_{rs}^P$ depend on a single non-negative parameter Λ:

$$\dot{\varepsilon}_{rs}^P = \Lambda\, g_{rs} \tag{8}$$

and, without loss of generality, take g_{rs} as a gradient of a function g(σ_{rs}), so that

$$\dot{\varepsilon}_{rs}^P = \Lambda \frac{\partial g}{\partial \sigma_{rs}} \tag{9}$$

Substituting Eq.(9) into Eq.(6) we may derive Λ and then finally $\dot{\varepsilon}_{rs}^P$:

$$\dot{\varepsilon}_{rs}^P = \frac{1}{H} \frac{\partial g}{\partial \sigma_{rs}} \frac{\partial f}{\partial \sigma_{ij}} \dot{\sigma}_{ij} \tag{10}$$

where:

$$H \equiv -\frac{\partial f}{\partial \psi_k} \frac{\partial \psi_k}{\partial \varepsilon_{rs}^P} \frac{\partial g}{\partial \sigma_{rs}} \tag{11}$$

is called hardening modulus. It is apparent that the continuity condition is fulfilled.

Eq.(10) is the well known equation for plastic strain rates used in the strain hardening theory of plasticity, which appears therefore as the simplest theory which is able to describe irreversible behaviour. The only hypotheses introduced so far are in fact

a) existence of an initial elastic domain, where only reversible strains take place

b) possibility of irreversible behaviour when the boundary of this domain is reached

c) evolution of such domain with plastic strains

d) dependence of plastic strain rates on a scalar parameter.

This implies that while the amount of plastic strain rates experienced depends on the amount and direction of $\dot{\sigma}_{ij}$, the direction of $\dot{\varepsilon}_{rs}^P$, i.e. the relative value of the independent components, depend only on the stress state σ_{ij} and not on the stress increment $\dot{\sigma}_{ij}$. This is at variance with the theory of elasticity, for instance, but is corroborated by some experimental evidence.

We may note that the functions that we should know to specify $\dot{\varepsilon}_{rs}^P$ are f, g and ψ_k. To derive also the total strain rates $\dot{\varepsilon}_{rs}^P$, we need further an elastic law. Clearly the ability of the model to reproduce the experimental behaviour under prescribed loading conditions will directly depend on the choice of such functions. The closest they will describe actual phenomena, e.g. the transition from the elastic regime to the elastoplastic, or the actual direction of the strain rates, the better the model will match experimental behaviour. Note that $\dot{\varepsilon}_{rs}^P$ depend on the derivatives of f and g. Therefore, an accurate choice of these functions is essential. From a geotechnical standpoint this is therefore the core of the problem.

In the following, some ideas will be given on what are believed to be the main requirements for such constitutive functions:

3.1 Initial elastic domain

The shape of the initial boundary of elastic behaviour should be similar to that of Fig.6. Strength in tension should be accounted for, in fact. Moreover, such a shape is appropriate for a variety of soft rocks and bonded soils, as shown by Nova (1992).

A consequence of such choice is that deviation from linearity, i.e. yielding for non collapsible geomaterials, occurs at lower deviatoric stress for high confining pressure than for moderate or low, see also Fig.3.

This fact has ample experimental support, see e.g. Pellegrino (1970), Allirot and Boehler (1979), Ohtsuki, Nishi, Okamoto and Tanaka (1981).

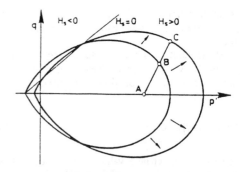

Figure 11. Evolution of elastic domain showing destructuration shifting (after Gens and Nova (1995)).

3.2 Subsequent yield loci, hardening and softening

The experimental determination of subsequent yield loci is more difficult. When the initial bonding is destroyed, the transition from elastic to elastoplastic regime is much less sharp and the exact position of the yield point is therefore difficult to assess. Moreover, many specimens should be tested to obtain the shape of the new yield locus. Since debonding may occur differently in different specimens, it is not likely that the experimental programme gives a clear trend of the subsequent yield loci. The choice of the evolution of the yield loci is a guess that should be verified (or better falsified) against experimental data concerning the overall behaviour.

If a specimen is confined under a sufficiently high pressure, it will yield at a point for which the hardening modulus H (given by Eq.11) is positive. Further strains will cause a reduction in H until failure will occur for $H = 0$, as can be retrieved from Eq.(10).

As a first approximation, for monotonic loading conditions, it is possible to assume that hardening is isotropic (the elastic domain changes in size but not in shape). Since however debonding causes soil properties degradation, with total loss of cohesion in the ultimate (large strains) state, it looks reasonable to assume that the hardening modulus is made up of two competing terms

$$H = H_s + H_d \qquad (12)$$

the first of which takes account of the enlargement of the elastic domain as for an unbonded soil, while the second always negative produces a shrinking due to the degradation. The evolution of the yield domain is as indicated in Fig.11.

It is clear from Eq.(12) that because of the degradation, H may become negative, the yield domain may consequently shrink and the strength may decrease. Such a phenomenon is usually called softening. Note that when H < 0 plastic strains take place even under fulfillment of Eq.(3). On the contrary Eq.(4) can never be fulfilled because it would imply an expansion of the yield locus and an increase in strength, in contradiction with the assumed negative value of H. Even without degradation, such a softening may occur under low confining pressures, as it is clear from Fig.4 when the confining pressure is low.

If the material dilates under shear even at high confining pressures, the term H_s should be assumed to depend on both volumetric and deviatoric plastic strains. At limit state (asymptotic failure) in fact, the rate of change of the hardening parameters ψ_k is zero. In particular for a single hardening parameter as for isotropic hardening

$$\dot{p}_c = \frac{\partial \dot{p}_c}{\partial v^p} \dot{v}^p + \frac{\partial p_c}{\partial e_{ij}^p} \dot{e}_{ij}^p = 0 \qquad (13)$$

where v is the volumetric strain and e_{ij} is the strain deviator. If p_c would depend on v^p only, as in the Cam Clay Model, for instance, at limit state no volumetric change could occur. In order to model the observed dilation, it is therefore mandatory that hardening depends also on deviatoric plastic strains.

3.3 Flow rule

In metal plasticity it is customary to assume that $g \equiv f$ (normality rule). Experimental as well as theoretical considerations (Lade 1988, Nova 1991) suggest however that in general $g \neq f$ for sands and clays, even though in the latter case the deviation from normality is not as large as in the former one.

3.4 Elasticity

Even when the stress state is within the elastic domain, the behaviour of any geomaterial can be considered rigorously linear elastic only up to axial strains of the order of 10^{-5}. For larger strains, microslips or closing of microcracks occur, so that the observed behaviour is neither linear nor elastic. As a first approximation, however, linear elasticity can be assumed to be valid within the entire elastic domain.

Non-linear elasticity (for instance variable elastic moduli with isotropic stress level) can be accounted for, if necessary. This may cause problems, however, since it may not be simple to choose an appropriate elastic potential. Only the existence of such a function guarantees in fact that a non-linear elastic model predicts no energy generation in all possible closed stress cycles.

4 CONSTITUTIVE FUNCTIONS

In order to show that a model characterised by the structure presented above can describe the experimental behaviour of various types of soft rocks in a variety of different tests, a short presentation of a particular model will be given in this section. A more complete presentation of various aspects can be found in Nova (1992), Lagioia and Nova (1995) and Lagioia, Puzrin and Potts (1996).

Consider first the plastic potential. As shown in Fig.7, during the destructuration phase the ratio between the volumetric and the deviatoric strain-rates is constant. These are actually plastic strain rates since both the isotropic and the deviatoric stress are constant and consequently elastic strains are nil.

If we define the dilatancy d as,

$$d \equiv \frac{dv^p}{d\varepsilon^p} \equiv \frac{d\varepsilon_1^p + 2d\varepsilon_3^p}{\frac{2}{3}\left(d\varepsilon_1^p - d\varepsilon_3^p\right)} \quad (14)$$

and the stress ratio η as

$$\eta \equiv \frac{q}{p} \equiv \frac{\sigma_1 - \sigma_3}{\frac{1}{3}\left(\sigma_1 + 2\sigma_3\right)} \quad (15)$$

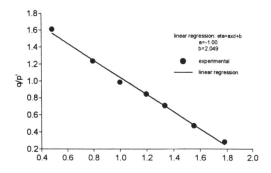

Figure 12. Stress ratio-dilatancy relationship during destructuration of Gravina calcarenite (after Lagioia and Nova (1995)).

as in the Cam-Clay model, it is possible to fit the data with a straight line:

$$\eta = M - \mu d \quad (16)$$

where M is the stress ratio for zero dilatancy and μ gives the slope, Fig.12.

Eq(16) was used by Nova and Wood (1979) for their model of sand. Such an equation has the defect that for isotropic loading ($\eta = 0$) the predicted dilatancy is finite, while, for symmetry reasons, it should tend to infinity. Nova and Wood (1979) assumed in fact another relation to be valid for low η values. Recently Lagioia, Puzrin and Potts (1996) proposed a more convenient relationship which is valid for low and high stress ratios:

$$d = \frac{M - \eta}{\mu}\left(\frac{\alpha M}{\eta} + 1\right) \quad (17)$$

For $\alpha = 0$ Eq(17) coincides with Eq(16). If $\alpha = 0$ and $\mu = 1$ the stress ratio-dilatancy relationship coincides with that used in the original Cam-Clay model (Schofield and Wroth 1968). Fig.13 shows the range of curves one can obtain as plastic potentials starting from Eq(17) for $\mu = 1.00001$ and various α values. For α close to zero the plastic potential coincides with Cam-Clay, while for larger values it can be made to be very close to the expressions given by Nova (1988) or Kim and Lade (1988).

The same expression, but in general with different parameters M_f, α_f, μ_f, can be used for the loading function. Indeed it is evident from Fig.13 that high α values give rise to curves which are very much similar to that fitting the yield points in Fig.6. Note that when the parameters characterising the plastic

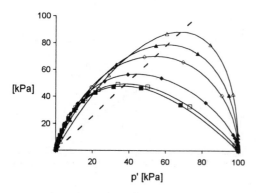

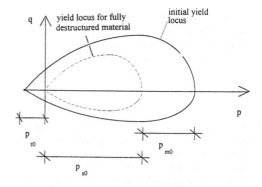

Figure 13. Plastic potentials in the 'triaxial plane' obtained integrating Eq(17) for $\mu=1.0001$ and various α (after Lagioia et al. (1996)).

Figure 14. Yield locus for a bonded material.

potential are different from those which characterise the loading function, the normality rule is violated. The loading function one can derive from Eq(17) (with parameters indexed f) is appropriate for granular materials such as sand. The bonding or cementation effect can be taken into account by introducing two new variables:

$$p^* \equiv p + p_t \tag{18}$$

$$\eta^* \equiv \frac{q}{p^*} \tag{19}$$

where p_t is a constitutive parameter closely connected to the tensile strength of the material.

By expressing the loading function in terms of p^* and η^*, the yield locus appears to be enlarged and shifted to the left, in the region of negative (tension) isotropic pressures, Fig.14.

For an intact material, the intercepts of the yield locus with the hydrostatic axis are ($-p_{t0}$, 0) and (p_{co}, 0), p_{t0} being the initial value of p_t. The parameter p_{co} can be split in two parts p_{so} and p_{mo}, the first equivalent to the isotropic preconsolidation pressure of a soil and the second linked to bonding. As the soil experiences plastic strains the hidden variables p_s, p_m, p_t change according to the following hardening rules:

$$p_s = p_{so} \exp\left[v^p + \xi \int \left(de_{ij}^p de_{ij}^p\right)^{\frac{1}{2}} \right] \tag{20}$$

$$p_m = p_{mo} \exp\left[-\rho_m \varepsilon_d^3\right] \tag{21}$$

$$p_t = p_{to} \exp\left[-\rho_t \varepsilon_d\right] \tag{22}$$

where:

$$\varepsilon_d = \int |dv^p| \tag{23}$$

The first equation is similar to that used for sand by Nova (1977) and Wilde (1977). The other two were proposed by Lagioia and Nova (1995) to describe the behaviour of Gravina calcarenite. In particular the cubic expression (21) allows to describe the occurrence of softening under purely hydrostatic loading, Fig.15. This phenomenon is linked to the crushing of the bonds and the consequent degradation of the rock which is transformed into a soil.

As there are three hidden variables the hardening modulus H can be expressed as

$$H = H_s + H_d = H_s + H_t + H_m \tag{24}$$

While H_s may be positive or negative depending on the state of stress, H_d, the degradation modulus, is always negative. The material may undergo overall hardening or softening depending on the relative value of H_s and H_d. For example Fig.16 shows the evolution of the yield locus during destructuration for Gravina calcarenite.

The generalisation to the full stress space is straightforward. The stress ratio is replaced by the second invariant of the stress ratio tensor:

$$\eta_{ij} \equiv \frac{s_{ij}}{p} \tag{25}$$

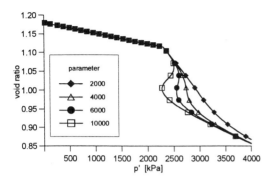

Figure 15. Void ratio change in hydrostatic test, showing the possibility of softening (after Lagioia and Nova (1995)).

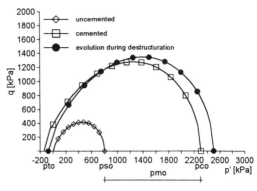

Figure 16. Evolution of the yield locus during destructuration (after Lagioia and Nova (1995)).

and the parameters M and M_f characterising the plastic potential and the loading function are made to depend on the Lode's angle θ:

$$\vartheta \equiv \frac{1}{3} arcsin\left\{-\frac{\sqrt{27}}{2}\frac{J_{3\eta}}{J_{2\eta}^{3/2}}\right\} \qquad (26)$$

$J_{2\eta}$ and $J_{3\eta}$ being the second and third invariant of the stress ratio tensor:

$$J_{2\eta} \equiv 1/2\ \eta_{ij}\ \eta_{ij} \qquad (27)$$

$$J_{3\eta} \equiv 1/3\ \eta_{ij}\ \eta_{jk}\ \eta_{ki} \qquad (28)$$

The functions $M(\theta)$, $M_f(\theta)$ are chosen in such a way that the sections on the deviatoric plane have a shape which coincides with the Matsuoka-Nakai (1982) failure criterion. A full mathematical presentation is given in Lagioia, Puzrin and Potts (1996).

5 COMPARISON BETWEEN EXPERIMENTAL DATA AND CALCULATED RESULTS

Fig.17 shows the experimental data and the model back predictions for three drained constant cell pressure tests on Gravina Calcarenite. Fig.18 shows the comparisons for undrained and p' constant test. In the undrained case the effective isotropic pressure is considered. Finally Fig.19 presents the theoretical and experimental results for an oedometric test. It is apparent that the model captures all the essential features of the material behaviour.

Fig.20,21 show the results of an isotropic

consolidation test and a series of drained tests on a different material, a marl of the Corinth Canal, data after Anagnostopoulos, Kalteziotis, Tsiambaos and Kavvadas (1991), showing no abrupt destructuration.

Fig.22,23 show the results for isotropically and K_0 consolidated undrained tests on Stevn's Klint chalk.

Other materials have been considered: artificial calcarenite (Coop and Atkinson 1993), tuff (Aversa, Evangelista and Ramondini 1991), oolitic limestone (Elliott and Brown 1985), grouted sand (Di Prisco, Matiotti and Nova 1992), and the quality of the results is similar. It is apparent therefore that the model can describe successfully the behaviour of an entire class of materials.

6 USE OF THE MODEL IN FINITE ELEMENT METHOD COMPUTATIONS

The model has been implemented in the finite element code ICFEP (Imperial College Finite Element Programme), and applied so far to the modelling of a load-settlement behaviour of a shallow strip foundation, 10 m. wide, on a layer of the Gravina calcarenite (Lagioia and Potts 1996). Fig.24a,b shows the contours of iso- p_t for a settlement of 25 and 250 mm, respectively. Since p_t continuously decreases with increasing plastic strains, it gives an idea of the degradation of the material. For settlement of 25 mm only a small zone below the edge of the foundation is destructured, while for 250 mm much of the material is degraded, the most intense region of destructuration being a narrow restricted zone below the edge. This is evidence of punching failure, what is confirmed by the pattern of

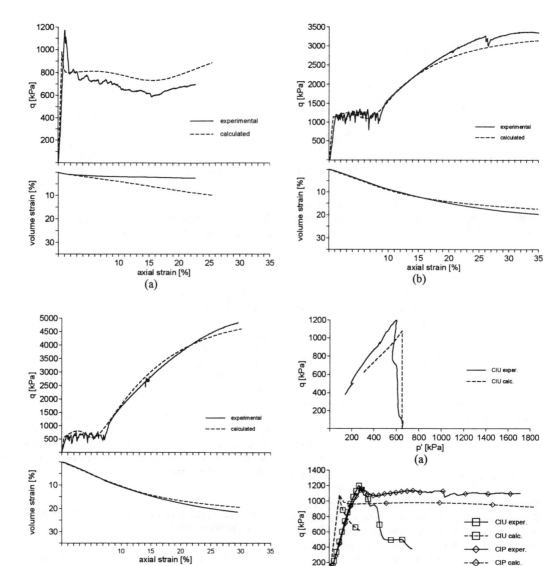

Figue 17. Drained constant cell pressure test on Gravina calcarenite. a) σ_3 = 200 Kpa, b) σ_3 = 1300 KPa, c) σ_3 = 2000 Kpa (after Lagioia and Nova (1995)).

Figure 18. Undrained and p' constant test on Gravina Calcarenite: a) undrained effective stress path, b) stress-strain relations in both tests (after Lagioia and Nova (1995)).

the rate of displacement vectors which are directed vertically below the foundation and are negligible elsewhere, Fig.25.

At present, experimental results for this type of problem are not available. Fig.26 shows however the predicted average pressure-displacement curve for the foundation and a set of pressure-settlement curves for 140 mm. diameter plate loading tests on

Lower Chalk at Cambridge (Lord 1989). The predicted bilinear behaviour for the calcarenite is observed experimentally on a similar material. Indeed, the type of behaviour of Fig.26 b is typical of grades III to V chalk [27].

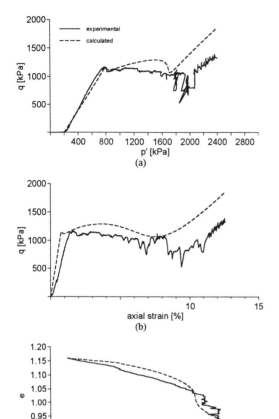

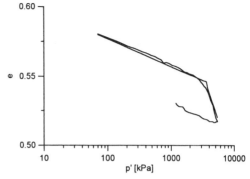

Figure 20. Hydrostatic test on Corinth Canal marl (data after Anagrastopoulos et al. (1991)).

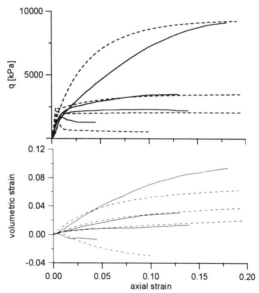

Figure 19. Oedometric test on Gravina calcarenite: a) stress path, b) stress-strain relation, c) void ratio change (after Lagioia and Nova (1995)).

Figure 21. Drained constant cell pressure tests on Corinth Canal marl (data after Anagrastopoulos et al. (1991)).

7 DIAGENESIS AND WEATHERING

The theory presented so far gives also a framework for the modelling of diagenesis (or artificial cementation such as grouting) and weathering. The size and shape of the yield loci depend in fact on the cement contact and the density of the material (Huang and Airey 1993). Consider Fig.27, which presents a series of yield loci at different void ratios (or porosities). The lower the void ratio the wider the elastic domain.

Assuming for the sake of simplicity that the shape does not change, the size of the yield locus can be related to the void ratio by a relation such as

$$e_o - e_o = \lambda' \ln \frac{p_{ci}}{p_{co}} \qquad (29)$$

which is similar to the consolidation equation for soils.

For instance, the value of λ' one can retrieve from the data of Huang and Airey for artificial calcarenite is 1.212. As a consequence, the yield deviatoric stress q_y increases with decreasing porosity, as

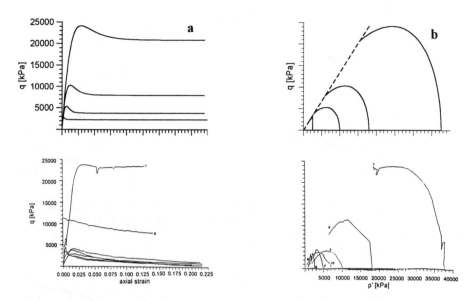

Figure 22. Isotropically consolidated undrained tests on Stevn's Klint chalk: a) stress-strain relation, b) effective stress paths data (after Leddra (1988)).

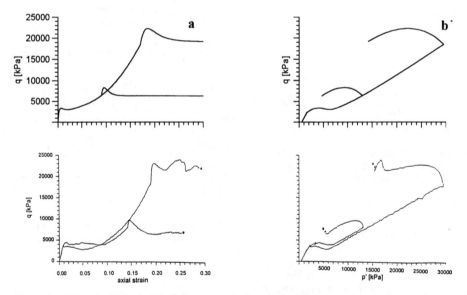

Figure 23. K_0 consolidated undrained tests on Stevn's Klint chalk: a) stress strain relation, b) effective stress paths (data after Leddra (1988)).

observed on chalk by Jones and Preston (1987) (fig.28).

The conceptual model of diagenesis is therefore the following. A sediment under its own weight is in a stress state represented by point A in Fig.29. The current yield locus (full line) passes through A and the origin of axes. Cementation causes a widening of the elastic domain adding p_t and p_m, while the stress state remains constant as depicted in Fig.29. Elastic properties change too.

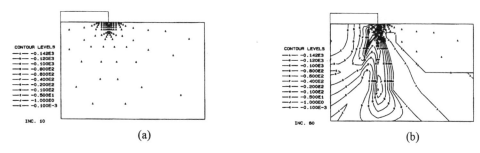

(a)   (b)

Figure 24. Contours of p_t under a strip foundation on Gravina Calcarenite: a) 25 mm. settlement, b) 250 mm. settlement (after Lagioia and Potts (1995)).

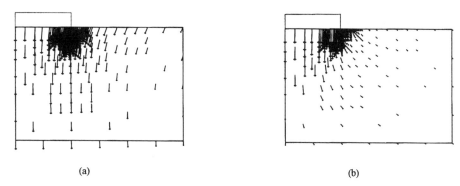

(a)   (b)

Figure 25. Rate of displacement vectors for the foundation of Fig.23: a) 25 mm. settlement, b) 250 mm. settlement (after Lagioia and Potts (1995)).

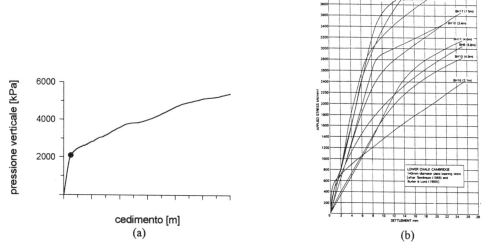

(a)   (b)

Figure 26. Load displacement curve: a) for the foundation of Fig.24 (Lagioia and Potts (1995)), b) measured for plate loading on Lower Chalk Cambridge (Lord (1989)).

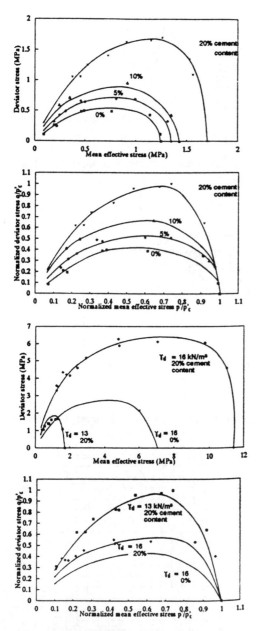

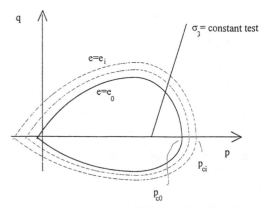

Figure 28. Variation of yield locus with void ratio.

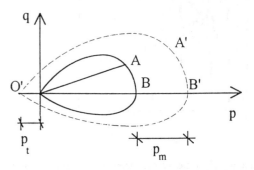

Figure 29. Schematic representation of the effect of diagenesis.

Figure 27. Yield loci for different voids ratios (after Huang and Airey (1993)).

Such a model can be used even for modelling the artificial grouting of an alluvial soil. A similar model was used in fact by Canetta, Cavagna and Nova (1996) to model the excavation of a tunnel in the alluvial soil of Milan, where an arch of soil around the tunnel was improved by grouting. It must be emphasised, however, that in that case it was necessary not only to change the material properties but even the state of stress, since grouting imposes anelastic strains, typically an increase in volume, which generate a field of self-stresses. Back calculations were in reasonable agreement with measured displacements at the surface and within the soil mass, but for the phase of grouting the predicted displacements were considerably lower than measured.

Conversely, the same conceptual structure can be used for modelling weathering. As weathering proceeds the porosity increases and the elastic domain shrinks (fig.30). Note in fact that even in the case there is no degradation of the bonds, the increase in voids causes an increase of the stress at the grain contacts so that the bond strength is

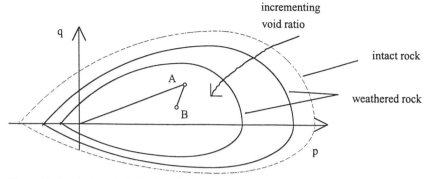

Figure 30. Evolution of elastic domain during weathering.

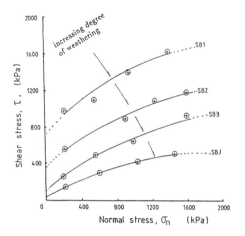

Figure 31. Shear strength envelopes for different degrees of weathering of granites (after Kimmance (1988)).

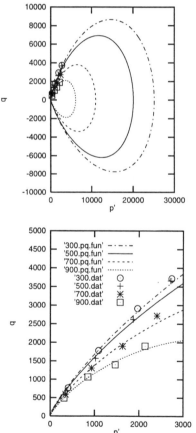

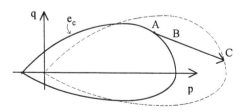

Figure 32. Yield locus and stress point (C) for a residual soil.

Figure 33. Yield loci for Westerly granite at different temperatures.

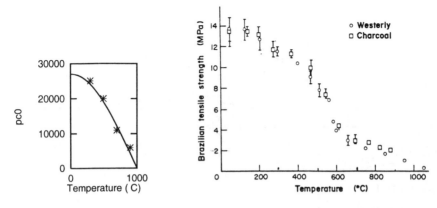

Figure 34. Effect of temperature on: a) calculated p_{co}, b) observed tensile strength (p_{to}).

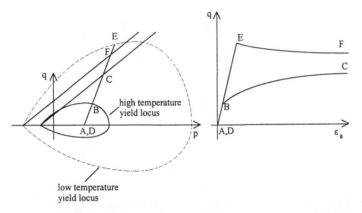

Figure 35. Conceptual modelling of the influence of temperature on brittle-ductile transition of granite: a)stress path and yield loci, b)stress-strain relationships.

reached for lower values of the stress applied at the boundary of the specimen. Fig.31 shows the variation of the shear-strength envelopes with increasing weathering for granite (Kimmance 1988).

For a certain degree of weathering and corresponding void ratio e_c, the stress state (point A in Fig.32) will be exactly on the yield locus. From that point onwards the elastic domain cannot shrink further. The stress point cannot lie outside the yield locus, in fact. By assuming that the vertical stress remains constant (actually there will be a small reduction due the decrease in density) the stress point will move from A to B in Fig.32 and correspondingly the yield locus will follow. Weathering causes therefore an initial increase in porosity followed by a decrease associated to collapse of pores. This qualitative result is in accordance with the trend of

data presented by Lumb (1962) on decomposed granite.

Full degradation transforms a granite into a residual soil, Fig.32, point C representing the stress state for full weathering. The tensile strength is nil at that point.

The effect of weathering on the elastic domain is similar to that of increasing temperature. Fig.33 shows the yield loci for Westerly granite at different temperatures. The data fitted starting from Eq.(17) are retrieved from Friedmann, Handin, Higgs and Lantz (1979). The calculated values of p_{co} as a function of temperature vary in a way similar to the tensile strength, Fig.34.

The conceptual modelling of the brittle ductile transition due to temperature increase is straightforward at this point. Under the same confining pressure a specimen of granite will yield at low deviatoric stress in the hardening regime for high temperatures, point B Fig.35, while it will fail in a brittle way under a much higher deviatoric stress (point E in Fig.35) for low temperatures.

8 CONCLUSIONS

The mechanical behaviour of soft rocks is quite complex: it is non-linear, dilative or contractive brittle or ductile depending on the level of confinement, it may show evidence of collapse of the pore structure. The mechanical model presented in this paper, which is an extension of the constitutive law used to model the behaviour of virgin soils, is able to describe the essential features of soft rocks behaviour.

A variety of materials was considered: natural and artificial calcarenite, tuff, marl, limestone, chalk. It has been further shown that the conceptual structure of such constitutive model can be used to describe diagenesis of sediments and the artificial grouting of alluvial soils. Conversely, the degradation due to weathering or temperature increase can be modelled in the same framework.

Strainhardening plasticity appears therefore as a convenient theory able to describe the behaviour of a wide class of geomaterials from soft clay to weathered granite. This is not only interesting from a speculative viewpoint but has remarkable practical consequences. Alternating layers of soils and soft rocks can be described by the same constitutive model characterised by different parameters. The modelling of grouting or rock degradation can be accounted for by change of parameters without the need of introducing new models. The preliminary numerical results obtained on calcarenite and grouted sand in practical boundary value problems are encouraging in this respect

REFERENCES

Addis, M.A. & M.E. Jones 1989. Mechanical behaviour and strain rate dependence of high porosity chalk. *Chalk, Proc.Int.Symp.* , Brighton: 239-244.

Allirot, D. & J.P. Boehler 1979. Evolutions de proprietés mécanique d'une roche stratificée sous pression de confinement. *Proc. 4th ECISRM, Montreurx*:15-22.

Anagnostopoulos, A.G., N. Kalteziotis, G.K. Tsiambaos, & M. Kavvadas 1991. Geotechnical proprieties of Corinth Canal marl. *Geotechnical and Geological Eng.*: 91-26.

Anagnostopoulos, A.G., F. Schlosser, N. Kalteziotis & R. Frank 1993. Geotechnical Engineering of Hard Soils-Soft Rocks. *Proc. Int. Symp.,Athens, Greece, 20-23 Sept 1993*. Rotterdam: Balkema.

Aversa, S., A. Evangelista, & M. Ramondini 1991. Snervamento e resistenza a rottura di un tufo a grana fine. *II Conv. Naz. CNR Geotecnica, Ravello*: 1-22.

Burland, J.B. & J.A. Lord 1969. The load deformation behaviour of Middle Chalk at Mundford, Norfolk: a comparison between full scale performance and in situ and laboratory measurements . *In situ investigation in Soils and Rocks, Proc.Conf., London 1969*: 3-5.

Canetta, G., B. Cavagna & R. Nova 1996. Experimental and numerical tests on the excavation of a railway tunnel in grouted soil in Milan. *Proc. Geotechnical aspects of underground construction in soft ground, London 1996.*

Coop, M.R. & J.H. Atkinson 1993. The mechanics of cemented carbonate sand. *Géotechnique*, 43(1): 53-67.

Di Prisco, C., R. Matiotti & R. Nova 1992. A mathematical model of grouted sand allowing for strength degradation. *Proc. NUMOG 4, Swansea 1992*: 25-35.

Dobereiner, L. & M.H. De Freitas 1986. Geotechnical properties of weak Sandstones. *Géotechnique* 36(1):79-94.

Elliott, G.H. & E.T. Brown 1985. Yield of soft high porosity rock. *Géotechnique* 35(4): 415-423.

Friedmann, M., J. Handin, N.G. Higgs & J.R. Lantz 1979. Strength and ductility of four dry rocks at low pressures and temperatures to partial melting. *Proc. 20th Symp. Rock Mech., Austin 1979*: 35-50.

Gens, A. & R. Nova 1993. Conceptual bases for a constitutive model for bonded soils and weak rocks. *Proc. Int. Symp. Geotechnical Engineering of Hard Soils-Soft Rocks, Athens 1993*: 485-494. Rotterdam: Balkema.

Huang, J.T. & D.W. Airey 1993. Effects of cement and density on an artificially cemented sand. *Geotechnical Engineering of Hard Soils-Soft Rocks, Athens 1993*: 553-560.

Jones, M.F. & R.H.I. Preston 1987. Introduction in: Deformation of Sediments and Sedimentary rocks. *Geol.Soc.Special Pub. 29, London*: 1-8.

Kim, M.K. & P.V. Lade 1988. Simple hardening constitutive model for frictional materials. I. Plastic potential function. *Computers and Geotechnics* 5: 307-324.

Kimmance, G.C. 1988. Computer aided risk analysis of open pit mine slopes in kaolin mined deposits. *Ph.D. Thesis* , Univ. of London.

Lade, P.V. 1988. Effects of voids and volume changes on the behaviour of frictional materials. *Int. J. .Num. Meth. Geomech.* 12: 351-370.

Lagioia, R. & R. Nova 1995. An experimental and theoretical study of the behaviour of a calcarenite in triaxial compression. *Géotechnique* 45(4):633-648.

Lagioia, R., A.M. Puzrin & D.M. Potts 1996. A new versatile expression for yield and plastic potential surfaces. *Computers & Geotechnics* 19(3): 171-191.

Lagioia, R. & D.M. Potts 1996. Analisi agli elementi finiti di una fondazione superficiale su di una calcarenite debolmente cementata. *Atti riunione Gruppo Naz. di Coord. per gli Studi di Ingegneria Geotecnica del C.N.R.*: 221-224.

Leddra, M.J. 1988. Deformation of Chalk through compaction and flow. *PhD Thesis*, University of London.

Lord, J.A. 1989. Foundations in chalk. *Chalk, Proc. Int. Symp., Brighton 1989:* 245-252.

Lumb, P. 1962. The properties of decomposed granite. *Géotechnique* 12 (3): 226-243.

Maccarini, M. 1967. Laboratory studies of weakly bonded artificial soil. *Ph. D. Thesis*, Univ. of London.

Matsuoka, H. & T. Nakai 1982. A new failure criterion for soils in three dimensional stress-space. *I.U.T.A.M. Conf. Deformation and Failure of Granular Materials, Delft 1982:* 253-263.

Nova, R. 1977. On the hardening of soils. *Archiw. Mech. Stos.* 29(3): 445-458.

Nova, R. 1988. Sinfonietta classica: an exercise on classical soil modelling. In A. Saada & G. Bianchini (eds.), *Proc. Int. Symp. Constitutive Equations for Granular Non-cohesive Materials, Cleveland 1987:* 501-520. Rotterdam: Balkema.

Nova, R. 1991. A note on sand liquefaction and soil stability. In C. Desai (ed.), *Proc. Int. Conf. Constitutive Laws for Engineering Materials, Tucson 1991:*153-156. ASME Press.

Nova, R. 1992. Mathematical modelling of natural and engineered geomaterials. *Europ. J.Mech. A/ Solids* 11, Special Iusse: 135-154.

Nova, R. & D.M. Wood 1979. A constitutive model for sand in triaxial compression. *Int. J. Num. Anal. Geomech.* 3(3): 255-278.

Ohtsuki, H., K. Nishi, T. Okamoto & S. Tanaka 1981. Time dependent characteristics of strength and deformation of a mudstone. *Proc. I.S. Weak Rock,* Tokyo 1981: 119-124.

Pellegrino, A. 1970. Mechanical behaviour of soft rocks under high stresses, *Proc. 2nd ICISRM,* Belgrado 1970.

Roscoe, K.H. & J.B. Burland 1968. On the generalised stress-strain behaviour of 'wet' clay. *Engineering Plasticity:* Cambridge Univ. Press, 535-609.

Schofield, A.N. & C.P. Wroth 1968. *Critical State Soil Mechanics:* Wiley.

Siwak, J.M., J. Prevost, G. Pecquer & A. Mikolajczak 1993. Behaviour of chalk. *Geotechnical Engineering of Hard Soils-Soft Rocks, Athens 1993:* 1657-1662.

Wilde, P. 1977. Two invariants depending model of granular media. *Archiw. Mech. Stos.* 19(4): 799-809.

Eurock '96, Barla (ed.) © 2000 Balkema, Rotterdam, ISBN 90 5809 843 6

Present understanding and predictive capability of the stability conditions for very high rock cuts

J.C. Sharp

Geo-Design, Jersey, UK

ABSTRACT: The prediction of the stability of very high rock cuts is a complex interactive subject involving many fundamental disciplines of rock mechanics. This paper provides a general overview of the procedures and processes currently used. In addition it attempts to examine in more detail complex mechanisms involving stress and structural interactions that lead to failure initiation. The role of progressive failure is also reviewed and emphasised. Brief reference is given to the issue of precedent behaviour. The principal conclusion reached is that a more formal reporting and evaluation of major slope design methodology and failure response are required to allow the subject to be addressed more rationally in the future.

1 INTRODUCTION

The invitation to provide this paper was made by Professor Giovanni Barla who showed great insight in the provision of the title. Of note were the omissions of the word *analysis* and also the slight limitation on the enormous subject matter allowed by the qualification *Very High* Rock Cuts.

The subject of rock slope stability is a vast area of rock mechanics that embodies most of the fundamental disciplines in terms of the geological characterisation and strength - deformability of rock masses, the understanding of external forces due to such factors as groundwater, seismicity and the important areas of control relating to excavation or cutting methodology as well as stabilisation techniques.

In this presentation a survey has been carried out of the critical factors influencing stability conditions and an attempt has been made to assess how these factors can be addressed in terms of engineering design methodology. Above all, the potential limitations of such methods have been carefully examined.

It would be impossible in this paper to trace in any meaningful way the development of rock slope engineering up to the present time although the major contributions of many well known people in this field such as Stini, Hoek, Bray, Londe, Goodman, Barton, Patton, Jennings, Voight and many others need to be recognised.

The major and fundamental issue with all rock slopes, whether natural or excavated, is the understanding of the geological characteristics that constitute the rock mass. The entire design and predictive process is closely linked in terms of scope and justification to this intrinsic understanding. Particular to this issue is the large scale distribution of natural discontinuities in the rock mass that provide the definition and framework for both the mechanical characterisation of the rock mass and the hydrogeological regime.

In terms of the subject matter under consideration, it is convenient to firstly classify the rock slopes in question. To do this certain key words (phrases) have been applied as follows:

1.1 *Excavated Rock Cuts*

Implying the use of excavation methods to form an initial rock cut in slightly weathered or fresh rock in which discontinuities will have a major and controlling influence. This may also include undercutting of existing natural slopes to form a composite slope.

1.2 *Very High Rock Cuts*

Classified here as rock cuts with an overall height in excess of about 100 m.

The subject matter has been deliberately limited to large scale rock slopes in generally unweathered rock where structure would normally be assumed to be the dominant factor.

1.3 Complex Failure Mechanisms

The slopes under consideration are assumed to fail in a complex behavioural manner (because of both geological and scale factors) that involves the interaction of geological structure, intact rock material, water and in situ stress. In this respect the potential for more simplified mechanisms such as planar or two-plane wedge sliding and simple toppling have not been considered since they are adequately covered in existing literature.

A further basic assumption that was made was the inference that the slopes in question are required for economic or other reasons to be as steep as possible and in this regard merit detailed consideration in terms of rock engineering investigation and assessment. For this reason particular emphasis has been given to more competent rather than weak rock masses.

The term *very large* tends to imply a reliance on the utilisation of the intrinsic stability of the rock slopes rather than the use of extensive engineering measures in the form of stabilisation. Where engineering stabilisation measures are however used in significant amounts it can be stated that the combined product of the intrinsic natural condition and the engineered measures must lead to a demonstrable and proven solution in terms of stability. (The adoption of extensive engineering measures can hardly be considered acceptable practice if failure may ultimately occur.)

2 UNCERTAINTY AND PHYSICAL PROCESSES

In rock engineering terms, the use of the word *complex* implies a degree of uncertainty that can be broadly addressed in two categories, namely:
- uncertainties related to individual parameter definitions (essentially geological, hydrogeological factors and strength data)
- uncertainties related to predictive or assessment mechanisms (behavioural models).

It is extremely important to address these two issues in a formal design sense and consider firstly the influence of parameter definition. In the past this has often been treated in a rather mechanical fashion by simply executing sensitivity analyses without due consideration of the rather grey and unconvincing spread of stability numbers that may be obtained with this approach.

In rock slope engineering, it is considered good and prudent practice to identify for a given geological environment the best physical interpretation of

conditions (discontinuity geometry, shear strength, etc.) prior to commencing any form of numerical simulation. It is unfortunate that too little emphasis is still given to the meaningful physical definition of the slope characteristics with emphasis on the rational treatment of geology. It is not uncommon that the form of analysis (and generally its limitations) will dictate the definition of input criteria thus denying a proper and fundamental account of the controlling physical process. Important in this field is the example set by Barton, Bandis and others in the physical description of discontinuity characteristics.

The next key step is to define (or attempt to) the physical process of failure in terms of a single or composite behavioural model relating to the key interacting characteristics of geology, deformability, strength and groundwater. The models used need not necessarily conform to available methods but should rather be a systematic examination of key factor relationships such as provided for in the Rock Engineering Systems approach. Complex problems will usually dictate the need for a degree of interpretation and innovation. The output from this stage should then identify a systematic design path that should be observed and define key parameters that need to be investigated, evaluated and assessed.

The importance of this critical evaluation stage in the overall design process cannot be overemphasised.

3 KEY PARAMETERS

Significant parameters that have an impact on stability conditions can be identified and briefly described as follows:

Geological Structure
The nature and geometry of the discontinuity pattern within the rock mass to be excavated.

Strength and Deformability of Discontinuities
The definition of key parameters of strength and stiffness for a given discontinuity stress state.

Rock Material State
The nature of the rock material and its potential change in state as a function of stress, strain and moisture.

In situ Stress State
The nature of stresses existing in situ and as potentially generated by the excavation state.

Groundwater Characteristics
The nature of the groundwater regime and its influence on rock mass strength.

Each of these key subject areas should be considered individually as behavioural models prior to attempting to combine them into a stability assessment.

It is considered particularly important to recognise the role of stress in very large rock cuts where the potential exists for the induced stresses to approach the strength of the rock mass in critical regions. This in turn can initiate deformation processes including the generation of induced fractures and hence failure mechanisms. Little is known about this subject area and highly subjective thinking and global assumptions often prevail.

4 POTENTIAL FAILURE MECHANISMS

Through an understanding of the potential state of the rock mass in the various parts of the slope, it is possible to identify potential modes of failure at least in concept for further study.

Of key importance in large excavations is the influence of progressive mechanisms that lead to a change in state of the rock mass as a result of such factors as displacements, groundwater interaction or time related phenomena, all of which can lead to the generation of incrementally more critical conditions. Path dependency is an essential factor of such processes that needs to be addressed if only in a qualitative sense.

By examining in detail likely mechanisms or at least identifying parameters of conceptual significance, it is then possible to identify whether or not general modelling procedures are available to allow a synthesis or interaction of the individual processes via an overall slope model to be achieved.

Such overall models normally take the form of a stress - strain - displacement prediction as an initial step to gain insight into the overall induced condition coupled, if appropriate (but as generally applies), with the groundwater (effective stress) state. This initial model is an extremely important stage in identifying the overall state of the slope prior to excavation and potential failure initiation. Such models should not be created with the intention of simulating failure according to a preconceived mechanism. Rather should be aimed at deriving an understanding of the in situ state and identifying key issues for further development.

From a study of the in situ state and its potential variation as a function of parameter uncertainties, it is then possible to identify potential governing

mechanisms of behaviour. The basic method of assessment is to examine the state of individual discontinuities in terms of stress and deformation thereby deriving an understanding of the rock mass state in terms of proximity or otherwise to localised failure generally termed *overstress*.

Similarly the failure state of the rock material, in terms of induced stress levels, can be assessed and broadly quantified in terms of stress - strength ratios.

5 INITIATION OF FAILURE

Skempton and other early workers in soil mechanics recognised the key mechanisms of progressive failure and sought to identify critical sections of slopes at which failure might initiate.

Early studies in the field of rock slope engineering also sought to identify where *failures* were likely to initiate but without appreciating or determining the fundamental mechanisms involved. Thus Hoek in reporting Barton's studies on large scale block simulations of slopes identified the importance of only small shear movements in the rock mass. It was also appreciated that the tension crack was a *result* or *signature* of slope deformation and not a fundamental process of failure initiation. Such phenomena were recognised as being only the start of a very complex progressive failure process about which very little (in the early 1970's) was known. After a quarter of a century it is important to examine whether the state-of-the-art has changed demonstrably.

It is intuitive that the key and focal mechanisms must be associated with the lower slope zone since it is only here that stresses (particularly recognising groundwater effects) are sufficient to generate a degree of *overstress* (see above) which will initiate the first stages of what may develop into an overall failure. It is perhaps unfortunate that the powers of observation so far in common existence do not allow us to examine the significant overstress processes (particularly in the early stage) in the toe zone. Discontinuity shear is perhaps the key factor in this. (To infer conditions from a remotely generated tensile accommodation feature in the upper slope already implies a significant degree of behavioural history that in turn may indicate already critical conditions in terms of the failure process.)

The identification and behaviour of both active (potentially overstressed) zones and passive (resistant) zones within discontinuous rock slopes is still an enigma awaiting logical solution. Work by Burland and others into such phenomena in the field of soil mechanics is of interest but not as yet conclusive.

It is important to try and understand the conditions under which a slope becomes active and which can lead to subsequent overstressing of the toe zone. The overall validity of this mechanism particularly in terms of induced stresses also needs to be examined. In general it is considered most likely that failure will initiate at the toe zone with the mid/upper slope zones only becoming active as a result of lost support. In such a mechanism, the gravitationally induced stresses from the mid and upper slope zones will be a key part of the initiation process.

6 RECOGNISED FAILURE MODES - ROLE OF GEOLOGICAL STRUCTURE

Failure modes for very high rock slopes may be complex but they will almost certainly involve rock mass structure to some degree in terms of failure geometry. It is therefore essential to consider the influence of structure as an intrinsic part of any failure mode.

Simple discontinuity geometries and their influence through planar and wedge failure initiation are well understood and will not be discussed in detail. They are usually termed kinematically admissible failures. A key issue arising with such failures however is the depth to which propagation of essentially planar shear displacements can occur. The approach by Goodman utilising *key block* principles tends to emphasise the importance of small block instability as a key to larger scale potential failures. By observing that these are more common, it is inferred that the occurrence of large scale failures, requiring significant joint continuity, is less likely. Again scale effects (joint length related to joint strength) are of critical importance. It should however be recognised that in spite of the general principles, large scale failures along major discontinuity planes (generally faults) can occur particularly when water pressures (and effective stresses) become more influential with increasing scale and depth.

Complex discontinuity geometries result in a very significant change in the necessary level of understanding of slope failures since they are usually classified as *kinetically inadmissible*. Such failures are not covered to any extent in standard texts on rock engineering and are commonly termed *unforeseeable* when they occur.

Some more complex mechanisms generally applicable to high, steep slopes (rather than extended shallow angle failures typical of weaker rocks) can be identified as follows:

I *Multiblock Failures*
Failures along multiple discontinuity planes with significant internal interblock shear.

II *Toppling*
A fundamental failure mode as a simple condition of over-turning but recognisably much more extensive as a mechanism (or sub mechanism) in large scale slope deformations. The mechanism is a particularly important consideration in the case of foliated rock masses (schists, phyllites, slates).

III *Compound Failures*
This term is used here to consider slope failure mechanisms involving a combination of discontinuity and rock material failure. The generation of induced fractures (material failure) is intrinsically related to the generation of adverse stresses and is a particularly important feature of slope toe zones.

Special cases of this failure involve the stability of long inclined slopes subparallel to bedding in sedimentary rock sequences. Failure through the stronger rock units associated with overstress along weaker basal planes is the usual mode observed.

IV *Kink Band Formation*
This failure state is based on classical structural geological principles associated with kink band formation. It is generally associated with foliated or highly to moderately jointed rock masses and has been studied in terms of behavioural models particularly by Archambault and Ladanyi.

Whilst the principle has been well illustrated in terms of failure of physical models under biaxial loading it has not been applied formally to particular large scale failures. Nevertheless the mechanisms implied are of fundamental importance in the understanding of large scale mechanisms.

V *Footwall Buckling Failures*
This term is applied specifically to so-called *footwall* slopes in bedded rocks characterised as in III above by an initial deformation mode that is related specifically to geological structure. In such failures that characteristically occur with prior bed deformation, buckling of the more competent surface slab occurs leading to under-cutting and failure of the entire slope.

7 EMPIRICAL OBSERVATIONS - STABLE SLOPE PROFILES

A classical empirical approach to the study of rock slope stability is the observation of stable and unstable slopes in different rock types with either high or steep characteristics or a combination of both. Studies by Hoek and Ross-Brown record much of the earlier work on this subject. Whilst the results are undoubtedly of interest as a general guideline, they cannot be applied with any certainty as a design tool because of the overriding influence of site specific features including structural geology, groundwater and stress all of which are intrinsically variable.

The indicated limits proposed by Hoek and Bray are of interest but also include many cases where the practical limits were achieved as a result of excavation technique rather than stability limits. Stability limits in practice can be probably best recognised by an appreciation of natural high rock faces in glaciated or water eroded terrain. In competent rock masses such as dolomitic limestone these processes can often lead to slopes over 300 m in height at angles of over 70 degrees.

The only logical application of empirical slope height - slope angle criteria is undoubtedly for a consistent rock mass condition (geology and structural orientation relative to the slope). Detailed exercises of this kind, that have been undertaken for specific rock types such as phyllites or slates can however indicate very different limiting angles for apparently similar slopes in practice. Such variations which may be up to 20 degrees for slopes of the order of 250 m high indicate overriding true nature of small scale geotechnical processes (mostly poorly understood) on the overall stability response.

8 GEOMORPHOLOGY AND TECTONIC HISTORY

The original landform associated with excavated slopes is often of very major significance in the stability assessment of excavated profiles especially where these result in oversteepening of the natural profile as commonly occurs in civil engineering applications.

As a general rule, the design and excavation of slopes in relatively young geological or geomorphological environments that already display significant relief (mountainous or hilly areas) should be treated with considerable respect and caution. (Never undercut young mountains is a basic lesson that seems to be timeless in this regard.)

Another major aspect of geomorphology and tectonic history is the in situ and induced stress fields that occurs prior to and after slope excavation. In situ stress fields in mountainous areas are notoriously variable and near impossible to measure or predict on a representative scale in spite of perhaps overall continental or sub continental trends that can be identified. This limit of information poses very severe restrictions in slope stability prediction for high slopes with complex mechanisms where the interactions of stresses and geological mechanisms will govern behaviour.

A general approach that is often used is to determine the gravitationally induced stresses that may occur as a result of excavation. For a slope being generated on a geomorphologically young plateau such an approach may be reasonable given sufficient scale to allow proper incorporation of boundary effects. For slope excavations in more complex geological and geometrical environments which have a complex stress path history due to either tectonic or erosional mechanisms, the initial stress state may be extremely difficult to predict with any degree of certainty.

This state of affairs underlines the fact that for very high slopes where stresses are a potentially critical issue, knowledge of the initial stress field is of fundamental importance. The intractable problems with prediction (in which theoretically one would need to simulate the geological and geometrical history - clearly an impossible task) leave little option but to undertake in situ measurements, however formidable this task may be in practice.

Knowledge of the initial state of stress is thus one of the key issues facing the predictive process for high and especially steep slopes. A non-exhaustive study of the literature has revealed very few cases of stress measurements involving such slopes. Where they have been executed, invariably in conjunction with containment studies for penstocks or pressure tunnels, they have usually only involved measurement of the minimal principal stress using the hydrofracture process rather than the full stress state essential for stability assessments.

A further key recommendation in slope stability evaluation for very large slopes is the execution of a geomorphological assessment that will serve to indicate the potential stress history of the particular situation. Even if no clearly defined tectonic or geometrical (erosion) sequence can be identified, the potential complexities and uncertainties can at least be formally identified.

A final interpretation following such studies should be made of the overall geological - geomorphological regime including the potential influence of gravitationally or tectonically induced stresses. This should be developed on the basis of comprehensive plans and sections and should be accompanied by models that serve to demonstrate the evolution and stress - deformation history of the current situation. Via such means both the stress conditions and the strength of rock mass can be judged rationally or otherwise identified as being of severe complexity - a useful end result in itself.

A further process that needs to be incorporated in complex slopes is the identification of formal spatial and continuity (persistence) relationships for discontinuities. This somewhat traditional approach incorporating the skills and virtues of geological observation and cartographic interpretation cannot be replaced by probabilistic analysis of somewhat *random* data which only confounds and obscures, rather than elucidates the solution.

9 STABILITY PREDICTION PROCESSES - INITIAL CONSIDERATIONS

In terms of large complex slopes there are no set guidelines or practices for the prediction of the stability state prior to and following excavation.

In general, as noted above, it is essential to undertake a comprehensive assessment of the current geological, geomorphological and hydrogeological setting to identify the structural, stress and groundwater state. This in turn implies an integrated and balanced approach of different specialists who can interact to provide a composite framework of understanding for the predictive process.

The physical understanding of conditions is the fundamental basis that provides confidence for any analytical process to predict the future stability state. It is of utmost importance to recognise the limitations of, and uncertainties inherent in, this knowledge if we are to attempt meaningful and reliable predictions.

It is relatively straightforward to identify the processes necessary to gain adequate knowledge (data) and understanding once the basic geological regime and potential rock mass characteristics have been identified. Translation of these into an analytical model requires experience and specialist skills based on a fundamental appreciation of the physical characteristics. Uncertainty resolution by appropriate assessment of staged predictions is an essential part of the entire modelling process.

10 STABILITY PREDICTION PROCESSES - SLOPE DESIGN APPROACHES

The need to understand both the initial and induced structural and stress states of the planned slope has been emphasised. Given adequate data on the structural geology it should be possible to construct a structural model whose response can be studied analytically. Additionally rational strength – deformability criteria for the discontinuities and the intact rock require to be defined.

The key factor then to be rationalised is the ambient stress field to be considered in the model. As noted previously this can vary from a quasi gravitational field achieved by model consolidation of the rock to a complex in situ condition derived from tectonic history.

The general design approach is then subject to the combined structural and strength model of the initial pre-excavation condition and achieves equilibrium. In itself, this process, which builds in stress history, is of extreme importance to the future predictive stability response. Always implicit in such modelling is the correct assumption of the effective stress state due to groundwater pressure.

The next general stage of the design process is to execute the excavation in a representative analytical form and analyse and evaluate the induced deformations, stresses and groundwater regime.

An evaluation of the induced conditions (assuming always that they are representative) is the first key issue in understanding the potential stability state.

The next key issue that arises is the identification of the potential mode of failure. This in itself implies the recognition of a particular mechanism or an awareness of the generation of a potentially complex, progressive model of failure behaviour.

11 ANALYTICAL METHODS

It is important to recognise that the analytical method should not restrict the design process by the need to make simplifying assumptions. The process of large rock cut design is intrinsically complex and needs to be supported by adequate analytical resources and methods.

A logical sequence of analytical method selection should be made to identify the following:

1 Justification of methodology based on available data inputs.
2. The possibility of using different analytical approaches to facilitate comparison of predictions and hence reliability.
3. The form of the analysis in terms of stress - deformation prediction or equilibrium assessment.

For major rock slopes, the following minimum facilities need to be incorporated in the adopted method:

1. Representation of all salient geological structure.
2. Rationalisation and incorporation of a realistic stress state.
3. Modelling of discontinuity deformation - strength criteria.
4. Modelling of rock material stress - strain – failure state.
5. Representation of groundwater conditions.
6. Consideration of dynamic response (seismic loading).

Based on currently available numerical codes, this implies the use of a discrete element code such as UDEC with available subroutines to correctly model the constituent component behaviour (i.e. discontinuity (UDEC - BB) response). Interactive coupling of slope deformation and hydraulic response is complex and heavily conditioned by uncertainty. Groundwater conditions logically therefore should constitute an analytical input, with uncertainty allowances, based on observed field response data.

Certain Finite Element routines which allow discrete displacements along joints to occur can also be used and may be helpful in terms of comparative calibration at least in the prediction of overall response.

It should be recognised that the execution of all such models is complex and requires input and experience on the analytical representation of physical processes. Such operations are not well supported by available rock engineering texts and must be recognised therefore as specialised. Also it is important to appreciate that the complex path dependency of the models generally does not allow a rational appreciation of the actual numerical procedure to be gained. As a result one is forced into a position of implicit trust in the model and its associated predictions since they cannot be cross checked or validated by other means.

Both the above types of analyses are capable of predicting a deformation-stress state for the rock mass comprised of discontinuity and material functions. They are not however equipped to study overall equilibrium and incorporate specific failure models.

The prediction of failure state is thus a complex issue which is not amenable to generalised definitions.

12 FAILURE STATE ASSESSMENT

As noted above, it is possible to identify different classes of failure that can influence stability processes. Where these are relatively simple (as for example for a well defined major wedge) they can be analysed in terms of a final equilibrium or failure state. However where the mechanism is complex and progressive equilibrium, as such, (unstable, meta-stable or stable) is either difficult or impossible to define.

The key issue for study is the identification of processes that can generate a failure path. These can be simply considered as:
1. Discontinuity displacement phenomena (singular or multiple).
2. Induced fracturing (failure of intact rock material).
3. Combinations of 1 + 2.

The term usually applied to discontinuity displacement phenomena that results in finite and irreversible behaviour is *overstress*. This should be related initially to specific discontinuities and can be considered as the state when the full strength of the discontinuity has been mobilised (under the effective stress state). It does not obviously imply overall failure unless the discontinuity has a simple relationship to the slope geometry that allows kinematic release to occur. Such overstressing phenomena are common in tectonic processes leading to shear zone and fault development. Clearly it is important to attempt modelling of the post peak strength - deformation state to understand its full potential effect.

The concept of discontinuity overstress can be extended to multiple discontinuities that may ultimately interact to generate a multiblock shear or shear - rotational failure zone.

The next component in the stability process is the induced fracturing (shear or shear - tensile - rotation mechanisms) of the so-called intact (block) material. Again this is a complex and interactive state with discontinuity behaviour.

Recognition of a rational failure state is clearly of extreme importance if one is to attempt to justify a particular degree of stability referenced to that particular state.

Unfortunately the selection of a failure state criterion is as (if not more) complex as (than) the identification of the failure mechanism itself and only general guidance can be given. For high, steep rock slopes reasonably safe design criteria could be considered as follows:
1. A recognisable and justifiable failure mode that can be reliably analysed.
2. General overall behaviour that is essentially elastic.
3. Localised overstress zones that are contained and do not lead to additional fracture generation or multiple discontinuity overstress.

4. For multiple overstress conditions that could generate a failure state utilise a design that restricts induced stresses to the order of 60% of failure levels or reinforce the slope to generate same conditions.

All the above conditions should be satisfied taking into account a representative ultimate critical loading induced by groundwater and dynamic effects.

13 FAILURE MODES

Whilst various different models of failure have been proposed above, their recognition in practice, and in particular the combined effects of different model types, poses significant restraints in terms of predictive behaviour.

Of note is the investigation, analysis and evaluation of key, critical rock slopes that have failed. A major concern in this regard is the mode of failure in terms of effect, speed and displaced volume. For high steep rock slopes above about 40 degrees in overall slope (but potentially less in some cases), catastrophic failure implying a rapid change in stability state or the release of significant energy can occur. Such events may be triggered by relatively minor or localised failure of the critical toe region of the slope but lead to gross, overall and sudden movements.

An attempt was made to define the type of failure in general terms in relation to catastrophic or non catastrophic events. Owing to the cause - effect situation that prevails, this in itself is problematic since only a small rockfall off a high slope could be disastrous in terms of a transportation corridor but acceptable in a mining environment. In engineering terms however, the design procedure needs to clearly recognise the effects of failure and be able to identify which conditions could prove catastrophic.

It is also worth noting that in the case of perhaps the three most notable rock slope failures at Vaiont (1963), Hope (1965) and Turtle Mountain, the cause and mechanism have remained in a state of debate and deliberation in spite of extended investigations by highly competent and experienced engineers.

To quote from Matthew and McTaggart in relation to the Hope Rockslide:

Apart from an earthquake, the causes of the 1965 landslide are not clear. The localisation of the failure surface over large areas within or at the margins of felsite sheets, and its general parallelism with the felsite sheets, with one of the conspicuous joint sets, and with the slight schistosity suggests all these may have contributed to the weakness ... The

role of water seems insignificant: consideration of meteorological conditions suggests that hydrostatic pressures were probably well below normal at the time ... Significantly, much loose debris persists on the slide surface, showing that the angle at which material would normally slide, even after rupture, is higher than the slope of the slide surface ... A major role in precipitating this landslide seems, by default, attributable to seismic activity. It cannot be assumed, however, that any earthquake of magnitude 3 or greater is capable of launching such a landslide; ...

A final, rather sobering, thought concerns the extreme difficulty that would have been involved in predicting the 1965 landslide ... Thus neither the surface of sliding nor the reason of failure could have been predicted even after a very thorough examination of the site. The study of stability of natural slopes such as this one clearly demands a humble approach.

Whilst these conclusions are related to the stability of a major natural slope, it is easy to accept the overall findings as being equally applicable to large, excavated rock slopes.

14 RISK ASSESSMENTS

Much slope engineering design and literature over the past 25 years have been concerned with risk assessment. The use of such an approach in terms of terrain or route evaluation to identify zones of relative risk is a valuable and promising area of development.

In terms of focussed slope stability assessments for a defined slope zone, the formal numerical application of risk assessment is extremely problematic and may introduce more confusion in terms of a meaningful end result. This inherently is a problem in terms of failure mechanism definition that, because of its imprecision in practice, cannot allow a meaningful risk analysis to be performed.

As a component of any study however, risk is an accepted phenomenon as well as the associated parameters of uncertainty and variability. All must be considered and accounted for at each and every stage of the design procedure.

15 EXCAVATION PRACTICE

Excavation methodology and practice are intrinsically linked to rock slope design particularly where steep excavation profiles are required. The practical generation of overall slope profiles steeper than about 40 degrees calls for the use of controlled excavation procedures that commonly employ pre-split or smooth blasting methods.

The design of individual excavation lifts or benches is also very important in terms of a sympathetic geometry relative to rock structure and to the slope design constraints. Of paramount importance in terms of bench design is the access facility that may be required for maintenance or the retention of minor localised failures. Bench height, face angle and width are all interrelated factors that need to be considered.

Feasible excavation practice is also important at the design stage both with regard to drilling and blasting technology and the ability of excavation plant to mechanically scale the excavated faces.

In certain cases of foliated or dipping sedimentary rocks, benches may or should be eliminated to prevent localised undercutting.

16 STABILISATION METHODS

Rock slope stabilisation for steep rock slopes is generally carried out using rock reinforcement in the form of anchors or bolts. Preferably these units should be tensioned and fully bonded to achieve maximum cost - benefit from the system. Design life in terms of corrosion protection is a further key issue.

Normally for very large slopes, stabilisation is only employed on a localised scale rather than to achieve an overall increase in stability. Examples exist with cuts of moderate height where long, high capacity rock anchors have been installed to produce a particular stabilisation result. In other cases, stabilisation has been applied over a large face area but only to a moderate depth to create stable benches for safety purposes.

Such methods need to be very carefully considered where rockfalls are a potential hazard since reinforcement and localised use of shotcrete facings or netting retention systems can be designed on an integrated basis to restrict or eliminate the rockfall source.

17 ROCKFALL PROTECTION

Slopes that are excavated to create a lower platform for public access (transportation corridors, residetial developments and the like) require special treatment in terms of rockfall protection. Of concern in the design of such slopes is the excavated profile which must be integrated and optimised in terms of the protection methods proposed.

In broad terms rockfall protection can be divided either into active face protection (rock reinforcement and netting over) or passive protection involving toe and possibly intermediate catch structures or control drape nets. The principal concerns relate to the slope height and geometry which will in turn dictate the likely trajectories and energy of rockfall events. To maximise safety, active or control systems that limit kinetic energy generation are normally preferred.

18 PRECEDENT PRACTICE - CASE EXAMPLES

Precedent practice in terms of rock slope engineering covers two main fields of application namely:
Existing natural slopes
Excavated slopes.

There is a great deal to be learned from the study of failures of both slope types in terms of the cause of failure, mechanisms and the derivation of predictive processes.

A study of large slope failures has shown a common trend of unexplained failure causes and, in complex conditions, a disparity of views between key causative factors such as structure, stress, groundwater etc. For major steep slopes where stress conditions in the toe region are a general concern, little specific evidence is presented as to the resultant failure mechanisms.

Many of the slopes concerned failed with little or no apparent warning. Failures that were recognised and monitored (notably Vaiont) were either hampered by limited observation or inadequate knowledge as to the potential adverse consequences of movement or water pressure changes.

The record of failures of high, steep excavated slopes is similar in scope and content to those that have occurred naturally. In some cases (notably Chuquicamata, Chile) monitoring was executed and the failure onset was predicted (albeit with a low level of understanding of actual mechanisms). In other cases, failures have occurred unexpectedly and even after significant subsequent investigation the cause of failure remains clouded and uncertain.

Unpublished data exist of major slopes that have moved in an accelerating manner and then unexpectedly slowed down even though causative, destabilising forces have not changed. In other instances slopes have behaved satisfactorily during adverse periods of precipitation and then experienced sudden unexpected movements when groundwater conditions were more favourable and when stability conditions would have been expected to have been enhanced.

In spite of this complex and unpredictable state of affairs, extremely valuable lessons and design guidelines are still to be learnt from failure records and inferred modes of behaviour. In addition general expectations of overall stability can be estimated by studying and comparing data on different slopes and environments.

19 CASE EXAMPLES OF ROCK SLOPE ENGINEERING INVOLVING COMPLEX FAILURE MODES

Case examples of rock slope design involving complex failure modes as reported in published literature are relatively limited. In general the major excavated rock slope designs and performance studies are related to the mining industry. Valuable lessons are also available from the study of natural slope failures and a limited number of civil engineering projects involving dam foundations and highway cuts. A common observation, as already noted in this paper is that design assessments of rock slopes involving complex failures are relatively limited. In particular the available data often preclude the generation of meaningful, comprehensive, behavioural models that allow stability levels to be derived.

Within the scope of this overview paper it is not possible to review the limited cases that are available. Instead a list of key slope evaluations, or in some cases failures, are provided below. The examples are chosen based on the author's own experience of complex failures or deformations of large slopes. It is to be hoped that other researchers in the field will develop such case histories into a more generic study which will also include and address appropriate analytical methodology and results. Only through careful assessment and evaluation of case studies will current predictive capability be enhanced. For each example keywords or a brief description are provided when relevant.

Vaiont, Italy
Failure of Mount Toc into hydro-electric scheme reservoir in the 1960's with catastrophic results. Key issues: failure surface continuity and nature, hydrogeology, reservoir fluctuations.

Rio Tinto, Spain
Partial failures and deformations of steep excavated mine slopes principally in period 1900 - 1920. Key

issues: regional stress state rock mass foliation, extensive rearward deformation zone, complex deep seated deformation zone.

Aznalcollar, Spain
Significant deformations and failures of mine slopes in foliated rock mass in period 1987 - 1996. Key issues: regional stress state (?), lack of obvious geological controls on slope responses, progressive failure response with variable and unpredictable relationship to groundwater response.

Hope Slide, Canada
Major natural slope failure at Hope BC in 1965. (See Section 13 above.)

Turtle Mountain Slide, Canada
Major catastrophic failure of a natural slope in dipping, sedimentary rock mass influenced in part by underground mining of the toe zone.

Jeffrey North Wall, Canada
Extensive study during the 1970's and 80's of the performance of a 300 m high excavated mine slope, partially reinforced and stabilised by drainage. Key issues: complex observed response of foliated rock mass with discrete movements at depth, extensive back analysis and simulation using FE and UDEC approaches, water pressure - movement correlation, lack of definable failure mechanism.

Chuquicamata, Chile
Failure of major mine slope with prior monitoring predictions but limited appreciation of mechanism. Ongoing stability evaluation programme to present day.

Southern Peru Copper, Peru
Development and assessment of major slopes at the Toquepala and Cuajone mines in excess of 500 m. Key issues: complex geological conditions, active tectonic region.

20 CONCLUSIONS AND DEVELOPMENTS

This paper has endeavoured to indicate the current status of rock slope design procedures applied to high, complex slope conditions. The general conclusions that can be drawn are given below in an attempt to focus attention on areas requiring significant technical development.
1. The design of major excavated slopes in rocks must recognise the complexity of the process as evidenced by the many unexplained failures that

have occurred both for natural and excavated slopes.

2. The design process is in general limited by the available data. Even with comprehensive, at depth investigations major uncertainties still prevail. The art of developing, evaluating and rationalising geotechnical data in a form suitable for engineering design input needs to be pursued and recognised.

3. The limited understanding of large scale failure mechanisms particularly in terms of stress - structure interaction phenomena and complex deformational modes (including rotation) needs to be continually addressed and improved. Again this calls for well defined physical models and careful consideration of the unexpected.

4. Given adequate physical definition, analytical models can be utilised to further our understanding of the complex interactive behaviour of large slopes. The findings of such analyses should only be considered as a design aid and not an end in themselves.

5. The singular use of risk or probabilistic analyses is not favoured for high slopes subject to complex failure modes.

6. The comparison and balance of predictive design with precedent practice is an essential element of rock slope design. If lack of precedence exists or uncertainties are identified, due care, caution and judgement must always be exercised.

7. To validate design predictions, monitoring should always be considered along with trial slope sections to build up reliable and specific precedent for the particular project. This is common practice in mining but could be further applied in terms of civil engineering applications during the construction period in order to determine the final safe profile or required level of stabilisation measures.

8. The use of overall stabilisation measures for high rock cuts inevitably requires the use of extensive rock reinforcement measures. The design of such measures can be considered as relatively straightforward provided that the failure mechanism and stability state can be clearly defined. In the case of poorly understood or ill-defined stability conditions, such measures should be adopted with significant caution.

9. The use of drainage (pressure dissipation) measures to enhance rock slope stability where adverse groundwater conditions exist can be considered as generally worthwhile. More data on the successful use of such measures would be of benefit to understanding the role of drainage in practice.

21 ACKNOWLEDGEMENTS

This paper could not have been written without the benefit of significant input through discussions from a number of the author's colleagues spanning a period of some 25 years of applied rock slope engineering. Particular mention must be made in terms of the inspiration provided by Dr. Stavros Bandis, Dr. Giovanni Barla, Dr. Laurie Richards, Dr. Oscar Steffen, Dr. John Byrne, David Pentz, Professor Richard Goodman and Peter Stacey. The willingness of several major mining companies to foster technical development and understanding of complex slope stability problems also needs to be recognised and recorded.

Eurock '96, Barla (ed.)© 2000 Balkema, Rotterdam, ISBN 90 5809 843 6

Specific and compared difficulties in works at great depth in mining civil engineering and oil industry from a rock mechanics perspective*

V. Maury

Elf Aquitaine, Pau, France (Presently: Consultant, Idron, France)

ABSTRACT

Great difficulties - unexpected rock ruptures and excessive deformation, sudden dynamic loading, delayed ruptures, influx of solids - are encountered when working at great depth, due to very high level of stress, temperature, pore pressure and fluid or temperature gradients around excavations. Accesses to workings are sometimes impossible and observations difficult as in oil or gas production. Consequences can be considerable in terms of safety and cost.

Very severe conditions met at great depth involve failure mechanisms not perfectly known, identified, and conditions of onset not correctly appraised. Moreover failure mechanisms may combine, making the analyses of the failure causes more difficult, and the remedies more or less efficient and uncertain.

While excavations for mining and civil engineering are accessible (except passes for safety reasons in mines, and pressure tunnels for dams), pressure and temperature may be very variable in oil/gas wells, and observations have to be done by means of logging tools.

It is sometimes claimed that conditions and problems met in mining and civil engineering are too different from those met for oil/gas production to be compared. We believe instead that these very different conditions and geometry of both types of excavation permits to collect a lot of information profitable to all the specialists. This is particularly true at great depth where rock mechanics environment is determinant. An exchange of information, observations, between specialists involved in projects of excavations or operations made with totally different purposes is highly fruitful.

This presentation is a very preliminary comparison and basis for a further exchange of experience between specialists and responsible persons involved in mining, civil engineering and oil or gas production at great depth.

After comparison of similarities and differences between requirements of each industry and effects of rock parameters, the impact of the Rock mechanics approach in terms of safety and economy is recalled. Some remaining difficulties and still unsolved issues are reviewed and some recommendations - or more modestly suggestions to help solving these problems - are made.

1 INTRODUCTION

Rock failures (or excessive deformation) - more generally unexpected behaviour - occurring at great depth are of paramount interest, as a compensation for their negative impact on the operations. The high level of stresses generated around the excavations for instance results in clear ruptures which can be compared to the existing criteria of failure. These criteria can in return be validated or contradicted, or some limitations can be recognised before a more general utilisation.

Moreover, rock failure comparison between those observed in various industrial activities are of utmost interest. Excavations made for mining or civil engineering are most of the time at atmospheric pressure with ruptures due to stress concentration which can be directly observed and monitored before any permanent lining is installed (water pressure tunnel for instance). Excavations made in the oil

* This keynote lecture was established with the help of D. Stacey from Steffen Robertson and Kirsten, Consulting Engineers, Illovo Johannesburg, South Africa & A.-P. Bois from Simecsol Group, Le Plessis-Robinson, France (presently: A.M.S.A. Technologies, Tassin la Demi-Lune, France) who are warmly thanked for their input.

industry (exploration or production wells) are submitted to various internal conditions (pressure, temperature, fluid and temperature gradients). They are much smaller and exhibit more complex rupture modes and failure mechanisms, but cannot be observed directly. Safety requirements are also totally different in mining works, civil engineering, and oil excavations. It is therefore profitable to compare what is done and observed in these various fields, and what can be learnt from one domain and transposed or adapted to another.

The objective of such an approach is certainly ambitious (maybe too much !) and the presentation given here has to be considered as a first step. This approach would surely deserve the attention of a working group of the Rock Mechanics Community, so enormous is the task to gather information coming from so separate and poorly connected fields of industrial activities. Rock mechanics is certainly the basic link between them.

In this first attempt, we had to limit to :

deep mining : South African gold mines are obviously at the first rank and have been solicited in this first step, but among others, deep mining in Coeur d'Alène Id. (U.-S.) and Kolar district (India) must not be forgotten. Conditions similar to mining at great depth can also be found at shallower depth when rock strength is low when compared to in situ stress for the considered depth

deep civil engineering : here are mainly the tunnels crossing mountain chains under very high overburden, and chambers made for various purposes (ventilation, etc...). Intermediate accesses or ventilation can imply the construction of vertical or inclined shafts. Some chambers for underground power plants are also in these conditions

oil/gas exploration and production: there are mainly the wells of any geometry made for any purposes and conditions.

As regard mining excavations and in order to draw conclusions from forms of excavations which can be compared, we will consider only observations made in permanent excavations such as shafts or passes (vertical), and tunnels (horizontal) and chambers (hoist, pump, crusher, refrigeration, workshop). Stoping operations and resulting excavations are not included since their associated difficulties are particular to the type of stoping carried out, and are not considered to be comparable with either civil or oil industry workings and difficulties.

As regards deep boreholes, boreholes made for scientific purposes must also be mentioned (Kola,

Sarty CEI, and KTB Germany), where a lot of observations and measurements could be done; they should be included in such comparison and would have to be added to this first step.

We will exclude also - in spite of a considerable interest - comparison of the overall behaviour of the rock mass such as an ore deposit submitted to ore extraction and an oil reservoir submitted to depletion induced by oil or gas production.

2 COMPARED CONDITIONS IN (DEEP) MINING, CIVIL ENGINEERING, AND OIL INDUSTRY

The general conditions encountered in these industrial domains are summarised in Table 1 and the main types of induced difficulties are indicated in Table 2. Some comments have to be added to these tables.

Effect of the internal static pressure (atmospheric - considered as nil - in mining and civil engineering and variable in oil production)

Mining has to build various types of excavations : vertical, namely shafts and passes, horizontal (tunnels), and chambers which have considerable horizontal and vertical extent (several tens of meters). A comprehensive review of conditions met in deep mining can be found in [1], [2]. To access deep level ore bodies, shafts are required for man access, hoisting and ventilation. Development in winding and rope technology has allowed single drops approaching 3000 m to be implemented, and sub-vertical shafts are being developed to in excess of 4000 m below surface. Shafts and rock passes can be built by conventional method, raise-bored or « shaft drilled ». The size of these excavations is commonly of several meters but can reach more than 10 m in some cases.

Civil engineering has to deal with horizontal excavations for main tunnels, but also with inclined or vertical shafts for secondary access, ventilation. Large chambers have also to be built for intermediate station, ventilation, pump-rooms specially for very long tunnels (> 10 to 15 km).

Man access to the working requires to be at atmospheric pressure, which implies a nil radial stress just behind the face after excavation.

In oil/gas production, if air-drilling has been occasionally developed (down to 4500 m in Aquitaine[3]), wells are drilled with a mud imposing an hydraulic pressure through a thin deposit of solid particles called « mud-cake ». Drilling can be made with a large variety of mud (water based-mud, oil based mud, very light mud (foams) but also very heavy (S.G. up to 2.30). Pressure regimes are highly variable from drilling conditions (in excess of some bars or tens of bars versus hydrostatic pressure), pressure tests (called « Leak off tests »), production or injectivity tests, fracturing procedures. If the shape of the hole is circular (at least at the beginning), wear of the drill pipe, dynamic behaviour of the drill pipe and break outs can result in ovaloïd shapes, rather irregular. All kind of geometry are now made from vertical (down to more than 9000 m), deviated, horizontal, with long extended reach (more than 6000 m).

Main problems encountered in mining and civil engineering are (i) in sidewalls, spalling (spitting, slabbing, sometimes called « strain bursting), instability due to geological structure, namely pre-existing fractures and bedding joints, interaction between stress induced fractures and geological structure to form unstable block and wedges, (ii) at the front (shaft bottom or tunnel face), stress induced popping up (rupture in extension of the face) can occur. Papers about methods of design, construction, and case studies are given in the South African publications about rock passes[4], [5], [6], shafts[7], tunnels[8], support of tunnels[9], [10](see for that several other papers by Ortlepp which could not be all quoted here). Problems met in raise-boring techniques (up to a diameter of 6.2 m) which can be of interest also for civil engineering were described by Stacey[11]. Several examples were given in civil engineering of such damages and evaluation in tunnels under thick overburden (blocks of a few cubic meters violently ejected at the Mont-Blanc Tunnel)[12],[13].

The extent of damages by spalling or strain bursting can reach 4 to 5 times the diameter of the openings. Instability of blocks and wedges, particularly in rock passes and shafts can reach impressive dimensions (initial 2.5 m diameter hole opened up to about 9 m diameter).

Mining tunnels are often subjected to mining induced stress changes after they have been developed and supported. They can be subjected to rock burst loading due to sudden stress redistribution due to mining progress[14] ; this type of secondary loading is much more frequent in tunnels (the shafts are usually long distance from the mining area). In this way the situation of mining tunnels is very different from civil engineering. Another major difference is the understanding of what constitutes « failure ». In a mining context, failure of rock is perfectly acceptable as long as the serviceability of the tunnel is not compromised. Conversely, civil engineering design and construction usually aims at preventing any

failure. At great depth this is usually an impossible task. One has to let the failure occur before permanent support is installed (strainbursting, squeezing, fault zone).

The situation is more complex in oil/gas wells, due to the existence of a radial stress which can be minimum, intermediate or maximum principal stress. Several types of instability and rupture modes can be predicted and observed : A, B, C shear mode, extension modes, and traction described in[15],[16],[17]. Spalling of the borehole sidewalls can be detected by the occurrence of cavings carried to the surface by the mud circulation, consequent problems can develop in re-entering the hole (problem to run in hole RIH) or pulling out the drill string (problems to pull out of hole POOH) without any warning or precursory signs of unstability ; losses of mud can at last be observed. Mud weight increase has for a long time been considered as the only remedy to problem stability, but Rock Mechanics analysis has shown that this remedy is efficient in vertical stress field, and possibly very detrimental in a low horizontal stress field and in the case of mud filtration in faults or fractures. Field examples have confirmed the theoretical approach and special equipment was proposed to mitigate this kind of problem.

Effect of temperature variation

In mining and civil engineering, temperature variations are limited to those induced by excavation and ventilation. At great depth (range 2000 to 4000m), virgin temperature can be in excess of 60°C and ventilation together with refrigeration must ensure a maximum temperature in working areas of no more than about 30°C.

In oil/gas production wells, particularly in the High pressure/High temperature environment (HP/HT corresponding to pore pressure up to 120 MPa and temperature of 190°C at 5000 m), cooling induced by mud circulation may induce cooling up to 50 to 60°C in the lower part of the open holes while at the same time the upper part is heated up to 40 to 50°C (average range of values).

In terms of stress at the wall, cooling (or heating) results in a decrease (or increase) of tangential and axial stress. When the internal pressure of the well is low (corresponding mud weight in the range of S.G. 1 - 1.4 or 1.5), the maximum stress is tangential and critical, and cooling which acts as an increase of internal pressure (given by the mud weight) stabilise the hole. Similarly heating may act as an equivalent decrease of internal pressure and result in instabilities. Failure of upper parts of openholes and delayed failures of lower parts of openholes due to natural re-heating (when stopping mud circulation) of bottomhole were clearly identified and treated accordingly (see examples in the next section).

Drilling in HP/HT conditions requires very high mud weight to prevent gas blow out. Then the cooling that acts always as an increase of the internal pressure may therefore become unfavourable, inducing thermal fracturing and be responsible for the so-called alternated « gains and losses » mechanisms[18].

The question can be raised to know if some delayed failures observed in mining and tunnels could not be linked, like in oil/gas wells to a progressive re-heating when ventilating stopes for instance[19]. Indeed tunnels constructed in hot rocks may suffer tensile cracking in their lining as the temperature is lowered. Such a case was observed in the Mont-Blanc Tunnel.

Hydraulic (or gas) gradient around openings

Two very different situations can be met at great depth : (i) case of homogeneous rocks, (ii) case of heterogeneity's, mainly crossing of faults filled with weak gouge material, often more permeable, prone to large water and solid influxes.

In Mining, hydraulic gradient around excavations can be considered as constant at the first glance, corresponding to the normal hydraulic gradient between hydrostatic and atmospheric pressure. Nevertheless attention has to be drawn to situations observed in shallow mines and quarries worked using the room and pillars method. The analysis of several cases of spontaneous collapse in such tabular excavations led us to consider that these collapses have been produced by change in hydraulic gradient around excavations resulting in progressive increasing hydraulic gradient and full hydrostatic loading on the immediate roof of the excavations[20]. Since similar ruptures were reported by Parker in deeper mines in Michigan, possible effects of such fluid pressure gradient variation around mining excavations must be considered cautiously.

In Civil Engineering, the normal water influx corresponding to homogeneous zones is discharged by means of the drainage system. So the question is more raised in terms of water and gouge material influx in water-bearing fault zones, which gave rise to famous case histories (Litani, Lebanon, St-Anne Mont-Cenis, France among others). This major risk for Civil Engineering excavations has to be treated by adequate pre-reconnaissance, pre-drainage and pregrouting of such zones. Then pre-support is installed before excavating the zone, support being mainly steel ribs.

The excavation through such zones is of major importance as it slows down the works and may even stops them. For instance the Lötschberg Tunnel excavated at the beginning of the century has been deviated after a major collapse. Moreover recovery of a T.B.M. stuck under 2000 m of overburden is a very difficult task.

In oil/gas production wells, a problem of identical nature may occur, less critical for safety but sometimes critical for production. Oil/gas wells can be produced in open-holes (in cemented rocks), or equipped with various systems (perforated cemented liners, slotted liners, etc..) in weak and poorly cemented rocks (sands, silts, chalks). When too high fluid gradients are induced (combined with high depletion inducing a stress redistribution at the scale of the reservoir), ruptures are observed resulting in solid particles influx (in weak sandstones, siltstones, chalks) very unfavourable for downhole equipments and surface installations and can lead to loss of the wells. In spite of many theoretical studies, the complexity of stress redistribution around borehole, cimentation, perforation, and pressure variation needs to turn to empirical methods.

Support

Methods of support design and construction in mining and civil engineering have been addressed in a lot of papers (see corresponding chapters in Comprehensive Rock Engineering and Panet[21], and it is not the purpose of dealing here with this extensive topic.

A special situation has to be observed for oil/gas wells during drilling. During drilling of hydrocarbon reservoirs or permeable rocks, the thin solid particles deposit (« mudcake ») left by mud circulation plays a major role in transmission of internal pressure to borehole wall. If the mudcake is of very low permeability, the internal pressure can be considered as a fully efficient support pressure. If the mudcake is very permeable (or if there is no mudcake as sometimes believed in front of shales or very low porosity rocks), the gradient between the internal pressure in the borehole and the hydrostatic pressure of the reservoir is dissipated in the borehole walls without creating an efficient support at the walls. A cake efficiency has been introduced in several analytical and numerical models, addressing also the pressure variation induced by thermoporomechanical effects. But the knowledge of the mudcake efficiency in bottomhole conditions is one of the most critical questions in case history and failure analysis.

Later on during well completion and production phases, the openhole is cased and deformation/rupture can be observed due to ground movements.

Drilling fluid/Rock Physico chemical interaction

In these three following sections, special reactions of rocks are considered, sometimes of common origin but experienced by various industries in different conditions with different features. As regards this range of problems, a more complete and frequent exchange of experience would be one of the most interesting to promote, namely between civil engineering/mining and oil wells drilling experiences.

In Civil engineering and mining, problems are separated according to following categories:

(i) « squeezing » rocks, i.-e. rocks exhibiting time dependent deformation, but without chemical reaction involved. The basic mechanism is the stress concentration around the excavation inducing excessive shear stress, plastic deformation and creep. A review of recent studies about squeezing rocks can be found in[22]. Any kind of weak rocks may therefore behave as a squeezing rock but the most common and prone to squeezing are certainly marls, claystones and shales, evaporites (gypsum, low strength evaporites).

Another source of time dependent deformation can be the pore pressure reduction induced by the excavation. In low porosity rocks such as clays, shales, a time dependent deformation may appear which in fact is a consolidation process. But the deviatoric stress conditions can result in excessive deformation as well as deformation in surrounding rocks

(ii) « swelling rocks », i.-e. rocks exhibiting deformation due to chemical reaction. The reader will see in the works of the ISRM Commission devoted to this topic the main recent issues on this topic[23]. The main chemical reaction present in civil engineering is hydratation of anhydrite and pyrite (and any composite rocks such as marls, clays or shales containing even very low percentage of these minerals) combined with stress and pore pressure variation. The consequences can be very important as can be observed in the tunnels excavated in the Jura chain (France-Switzerland).

In oil/gas wells drilling and production, the different nature of mud available (water based-mud, oil based mud, and now a large variety of mud with chemical additives) enlarge the possibilities of rock/fluid interaction, but some problems are in fact also problems of squeezing and/or swelling rocks. We

will start here by the problems such as approached by the oil industry before reviewing some problems raised specifically by squeezing and swelling rocks.

Drilling fluid/rock interaction in oil/gas well drilling and production can be the sources of following problems :

(i) in poorly compacted clays (tertiary shales of Southern North Sea for instance), plastic behaviour which is not fully resolved by increasing the mud weight. A possible interaction due to ion exchange between clays or shales and water based mud is sometimes mitigated by use of Potassium Chloride (KCl). Oil based muds have also improved the situation but not always

(ii) in semi-hard shales, chemical exchanges are still possible between rock and mud and the situation has been substantially improved by using oil based muds. But ruptures by spalling of the walls giving a great quantity of « cavings » and break-outs is still observed. These instabilities which can be delayed and affect very thick layers of overburden are extremely costly. Their analyses require to identify first the main mechanism responsible of rupture : pure effect of structure (anisotropy of deformability and strength of the shales, stress reconcentration around the hole and internal pressure), chemical interaction of mud and rock, thermo-mechanical and thermoporo-mechanical effects (increase of pore pressure due to heating). Here also the use of KCl mud can improve the situation probably by changing the pore pressure in shales by a mechanism of ion exchange and a subsequent effect on the resulting effective stresses. Here the rock mechanics analysis is very complex but can be extremely profitable if some field data, of utmost importance can obtained.

(iii) in pure anhydrites, there are not always difficulties. It had been observed in mines (Billingham mine, U.-K.) and tunnels that very compact anhydrites are not sensitive to water (due probably to their low porosity). The question is different in marls or shales containing small percentage of anhydrite which can swell. The use of oil based mud is an obvious solution to these problems

(iv) the case of evaporites can show many different situations. In case of thick homogeneous evaporites (halite), there are sometimes no other solution than increasing the mud weight (drilling can be done with undersaturated brine, permanent dissolution compensating creep). In case of series of heterogeneous evaporites (such as Zechstein in Southern North Sea), - a very difficult situation -, alternation of soluble, very weak (bieschoffiete, carnallite, kieserite) and much stronger (polyhalite, dolomites) may at the same time make it impossible to find a satisfactory mud weight and brine saturation. Rupture, dissolution of some layers, creep, can result in huge cavings, inducing difficulties during drilling but also later after casing has been installed by brine circulation in damaged cementation. Extremely high strength casing can even be damaged in this way.

Squeezing rocks

We will limit consideration in the two following sections to the case of squeezing and swelling rocks as seen in civil engineering (and sometimes in mining). As recalled by Barla, squeezing is the time dependent large deformation which occurs around the tunnel and is associated with creep caused by an excessive shear stress. Deformation may continue over a long period of time. Several modes of ruptures or deformation can be observed[24,25], associated with deformation in squeezing rocks themselves. Many numerical and analytical methods were proposed to evaluate conditions of stability and support required[21,26].

Swelling rocks

Swelling due to hydration of anhydrite and sulfate rocks was recognised long time ago[27], many studies were made to improve the basic knowledge of the mechanisms involved[28,29] and procedures were proposed for design and construction[30,23].

Hydration of pyrites results in transient sulphuric acid then anhydrite and gypsum formation (if Ca ions are present) or other sulphates. This mechanism can be frequently observed on cores taken in the shaly overburden of hydrocarbons reservoirs. Hydration of pyrites (even 1 to 2% of pyrites are sufficient !) is produced first at the expense of the water contained in the sample and appears as an auto-desiccation. Growth of white sulfate crystals are observed, creating fine cracks in the sample which are not 'fissures de retrait' as commonly admitted (the overall length of the sample increases).

Nothing is known about the conditions of onset of such phenomenon in oil/gas well drilling. But a slight desequilibrium in oxydo-reduction potential of drilling fluid and the frequent presence of calcium leaves the question pending.

The question can be raised as to whether some case of swelling of clays observed in civil engineering are not due to this mechanism, and could have been ignored due to the low content of pyrites sufficient to induce the mechanism (most of the time this low content leads to their being considered as a secondary mineral unable to be active).

Dynamic loading

Three main cases can be found resulting in dynamic loading around excavations :

(i) in no seismically active zones, single openings (such as tunnels for civil engineering purposes) are not subjected to dynamic loading (the case of suddenly varying pressure in tunnels and oil wells is not considered here as a dynamic loading). (The case of rockburst occurring in tunnels in Norway can be questioned)

The most important case of dynamic loading of tunnels, galleries, shafts is certainly the case of dynamic loading induced by seismicity during mining operations. The reader will found a description and analysis of this crucial phenomenon in South African gold mines in[14], characteristics and features of the mechanism[31], effect on support design[32,33,34]. (See also the various Symposia and workshops devoted to this topic, Santiago Chile 1994, Kingston Ontario 1993).

Much less documented are the cases of induced seismicity by oil/gas production[35,36,37]. Practically nothing is known about the damages caused to wells and tubulars caused by this case of dynamic loading.

(ii) in active seismic zones, it has been noticed that very deep excavations are less sensitive to dynamic loading (due to natural seismicity) than shallower or surface installation (surface wave attenuation in depth). Critical parts are accesses of mines and tunnels (portals and zones under low overburden).

In active tectonic zones, some failure of wells were observed which were attributed to natural seismicity (Iraq, Iran). Nevertheless it is much more difficult to conclude in this case if this is the natural seismicity which has damaged the well or if we are in case of induced seismicity by oil operation or even gas leakage around cementation, such as described in[44].

(iii) an intermediate case can be found in highly stressed zones such as foothills or highly horizontally stressed zones. It can be questioned if some modification created by excavation (for instance injection of fluids under pressure in more permeable faults and fractures during oil/gas wells production cannot be responsible for induced seismicity and subsequent dynamic loading.

3 SAFETY AND ECONOMICAL IMPACTS OF ROCK MECHANICS APPROACH

It is worth illustrating with field examples the safety and economic impact of Rock Mechanics in some industrial fields. In spite of insufficient theoretical knowledge, analysis, and interpretation of the mechanisms involved, practical recommendations for construction, excavation, drilling or production procedures can be issued with significant consequences on economics and sometimes safety.

From mining, a Symposium was held in South Africa dedicated to the impact of rock engineering on mining and tunnelling economics (1991)[38]. Such an approach should be generalised and made in Civil Engineering, and Oil/Gas production. It is interesting to recall here some conclusions drawn by Wagner (in [38]):

« ...Rock engineering can make major contribution to the safety and economics of mining operations by creating favourable strata conditions which results in reduced losses and increased revenues... »

« Rock related accidents can be reduced significantly by training the workforce in basic strata control and support principles »

« The contribution of Rock Engineering to mining should be measured in terms of safety and economic parameters and not rock engineering parameters »

« The rock engineering profession on mines is unaware of the important contribution that it can make to the wellbeing of the industry »

« Present management structures on mines do not take full advantages of the potential contributions of rock engineering »

It is believed here that these conclusions apply also to civil engineering and oil production. With this aim in view, we gathered several examples related to drilling gas/oil wells, where the direct impact of recommendations for new wells could easily be evaluated by comparison. Some of these cases were briefly presented at the Opening Session of Eurock 94 (Delft) but not published. They were only summarised at the 14th World Petroleum Congress[39]. It was estimated advisable by colleagues from civil engineering and mining to recall the benefits of these field cases.

Determination of internal pressure (mud weight) in boreholes according to the in situ state of stress (Field Case N°1)

Two wells drilled in the early 80's in Southern Europe to a depth of about 2000m gave rise to repeated incidents of stuck pipe and cavings, in a very complex sequence of shales and sandstones. No solution could be found with mud weight increases, up to 1.98 and even 2, repeatedly tried without improvement. Three unsuccessful side-tracks resulted in 4 months of lost time (total cost of incidents related to both wells were estimated to 6 MUS$ in 1985).

A detailed analysis of incidents associated with a theoretical study was performed accounting for the regional state of stress. The result (summarised on sketch as proposed by Guenot) is that before reaching the mud weight (or internal pressure) corresponding to hydraulic fracturing, a tricky shear failure mode can be obtained at a mud weight of about 1.45 (depending obviously of the geomechanical properties of the rock)(See for further detail ref. [15, 16, 17] for the theoretical background of this diagram).

A recommendation was given, to not increase mud weight above this value (which was surprising for drillers and mud specialists), and to avoid any cause of internal overpressures such as swab and surge. Recommendations were also given for the shapes of tools to prefer, and drilling parameters. The new well was drilled without problem avoiding the 4 months of drilling rig time lost in the previous wells. Cost of the study, including a difficult field data analysis was in the range of 100 000US$, to compare to 6 MUS$ lost, meaning a "direct" return to the rock mechanics study for this well and field of 6/0.1 = 60.

Thermal effects

As previously mentioned, below a certain depth (around 1500m), cooling due to mud circulation can give a temporary stability due to the decrease of tangential and axial (compressive) stress. When the mud circulation stops, return to thermal equilibrium causes a re-heating of the bottom hole. This re-heating increases (tangential and axial) stresses which can exceed the failure criterion and be the cause of the postponed failure (well known by wireline loggers because it seems to occur after the 3[d] logging or the 15th hour).

At the same time, the upper part of the open-hole is heated by the mud circulation. This heating induces additional compressive tangential stresses which can cause failure. These failures occurring after some time were considered in the past as caused by creep. These were also observed in rocks not prone to creep and are caused by heating (these thermal stresses have been overlooked and underestimated in the past). Recognition of the role of thermal stresses has permitted the mud weight to be adjusted to counteract this effect, and results in a very slight and progressive increase of the mud weight with depth.

Although already observed in the past by some drillers, identification of this mechanism led to a complete evaluation of the thermal regime in the wells during drilling and of the induced stresses. It allowed the quantification of the slight and progressive mud weight necessary to mitigate ruptures of this origin and incidentally (in Aquitaine) the confirmation of the

0.01 s.g. mud weight increase per 100m of deepening so beneficial to get rid of troubles.

It is somewhat difficult to make an evaluation of the amount of money saved by methods deduced from this analysis. From people in the field involved in the last 12 wells in Aquitaine, which means a minimum evaluation ignoring the previous wells, an average of 12 to 15 rig days were lost per well implying a minimum of 140 rig days lost, say 10 MUS$. Again according to drillers in charge of operations themselves, ten to fifteen days of rig could be saved for each of the next 12 wells, representing at least 100 rig days (from 1985 till 1992). The cost of the study is difficult to evaluate because it came from a larger research project. A reasonable evaluation (including a part of the research project) would lead to a cost of 0.3MUS$. The "direct" return of this study, which is really a minimum by ignoring the time lost for previous wells, and limiting the "gain" to new wells of this field is of 7.1/0.3 = 24 at a minimum.

But the evaluation of the role of the thermal regime and its variation in the borehole led to new and important conclusions.

Awareness of the role of thermal effects in the borehole stability led to a study utilising a mud cooling system to improve stability[40]. In spite of unavoidable and systematic objections and scepticism when proposing something new, a pilot test was made and permitted confirmation of a lot of side advantages extremely beneficial to drilling conditions in terms of safety and economics. They were detailed in a recent SPE publication[40].

A better understanding of the wellbore geomechanical behaviour can thus lead to practical means of improving not only its stability but have important consequences on safety and economics. More than 70 wells have been drilled with this mud cooling equipment for a field development in the Gulf of Guinea. A second development in currently underway with the same type of equipment, mainly for the safety advantages. A direct quantification of the economic advantages is difficult, some drilling conditions being different from the past where the cooling systems were not yet installed. This equipment was also successfully used for 6 development wells in Aquitaine (France) and for a deep exploration well (6909m).

Drilling in shales

All of us are familiar with the difficulties met when drilling certain types of shales. Most of the time, the situation is improved by means of successive trials for the mud types and properties implying a very long and costly learning curve, not always valid for all of the well trajectories (deviations and azimuths), and giving results impossible to extrapolate. This sometimes degenerates into passionate debates between drillers exchanging results of apparently contradictory experiences (consult for instance two drillers of the same company about the influence of KCl for two shales in two different countries !)

Two field cases were reported in the recent years illustrating how a Rock Mechanics evaluation could help to shorten this learning curve and save money.

In the case of Nelson Field[41], a series of wellbore stability problems were experienced for exploration wells. The rock mechanics study undertaken resulted in a much better shale problem evaluation. From figures mentioned in the paper, it can be easily deduced that 70 days of rig time were saved (at the minimum) at a cost of about 110 000 US$ per day by means of this adequate rock mechanics evaluation of the wellbore stability, meaning an amount of 7.7 MUS$ saved for 5 wells (minimum). The direct cost of the wellbore stability study was in the range of 30 to 35 000 US$ but had taken advantages of a general shale research program and database undertaken by the Company costing several million dollars (but is considered as already amortised on other fields). Again the direct return of a study of 0.035 MUS$ to a 7.7 MUS$ (at least) saved is in the range of 7.7/0.035= 220.

Another case was reported to us by some members of the company, mentioning trouble costs averaging 19% of the cost of the well over 10 wells resulting in an estimated total 9MUS$ trouble cost, considered as typical for fields where detailed wellbore stability analyses have not been integrated into the drilling program.

A **case in North Sea[42]** describes the rock mechanics evaluation of cases where drilling problems generated by stuck drill strings accounting for 10 M£ per year. From recommendations given resulting from the study, cost of the stuck pipes were reduced to less than half resulting in 5M£ per year saved. Shell Expro counts that at least 30% (1.5M£/yr) of that can be credited to the rock mechanics study on borehole stability. The cost of the study was reported to us to be in the range of 250 000£, so the ratio of money saved for the next three years only (4.5M£) to the cost of the study (0.25 M£) is 18, and increasing for the coming years.

More interesting in the paper is the process used for the study : "...three main stuck pipe mechanisms were defined ..., The stuck pipe mechanisms are not the root causes. They can be further separated into root cause areas such as ...". More important yet :"

Every time a drillpipe became stuck, an "autopsy" of the incident was performed ..., the drill crew was interviewed. All efforts were made to draw a precise picture of the decisions and actions leading to the stuck pipe incident."

We will come back in the conclusions on this very important aspect of the analysis because we consider that this part of the study - the analysis of the field data - is an essential part of the failure mechanism analysis, sometimes difficult, but at least as important as any theoretical model development.

Mechanisms of instability due to fractures, joints, and faults movements

In cases of large subsidence such as in the North Sea, it is obvious that tubulars (casing and tubing) can be deformed and collapsed by ground movements and displacements. There are nevertheless other cases where casing and tubing collapsed without any noticeable movements at the surface, and are much more difficult to understand.

Three deep gas wells were lost in France (Aquiline), either during completion, work-oversee, or production, resulting in a loss of 250 000 MFF (40 000 MUS$), and a leakage of gas with H$_2$S at the surface, which means a serious safety concern for people in charge of operations. No seismic event, corrosion, or wear could be invoked as a cause for these failures.

A complete analysis[43],[44] brought two important results :

- these collapses were due to a sequence of circumstances : drilling of the reservoir with total controlled losses, damage and microannuli in the cementation, channelling of overpressured fluids, pore pressure increase in the upper level faults or joints triggering lateral shifts by shear stress release, slight deformation of the casings weakening them, and collapse in the case of new internal pressure variations (work over or production incident).

Accounting for this mechanism allowed for a corrected well architecture (as well as work over procedures) for the three new replacement wells which could be drilled, completed and are presently working and in good conditions. The design of all the wells in the field has now been modified accordingly.

Again the cost of the 3 lost wells (40 MUS$) can be compared to the cost of the study (0.2 MUS$) giving a direct return of 200 times the cost of the study, of the same order of magnitude as previously found in different circumstances.

- but more important, it appeared that the same mechanism could affect open holes during drilling : when the mud has insufficient sealing capacities and the mud cake is not fully efficient, filtration of the mud in fractures, faults, bedding joints more permeable than the rock matrix increase the pore pressure in these discontinuities and acts like a flat jack. Then the normal effective stress is released and the shear stress causes a lateral shift of the borehole axis. Once recognised, this displacement could be observed with logging tools such as the new borehole imagery tools (BHTV, FMS, USI)(see ref. [40]).

Identification of this mechanism is outstanding : it appears very common, and explains some abnormal behaviour of wells unexplained up to now : without any cavings or signs of matrix failure in the borehole. When the movements occur in a part of the hole already drilled, the driller is faced with problems of running in hole (R.I.H.), tight hole, abnormal torque (perhaps drillstring failure if the lateral displacement is important) and has to be solved by reaming. If the movements occur behind the tool (higher in the open hole), the driller is faced with the problem of pulling out of hole (P.O.O.H.), sometimes stuck in hole, with jarring perfectly inefficient in case of noticeable movements. It must be pointed out that these kinds of problems are in no case solved by a mud weight increase which enhances and triggers new movements. A possibility of back-reaming has to be foreseen, which can be done sometimes without great problem.

Note that a case of a lateral shift of 3" of a hole of 8"1/2 diameter has been reported to us which obviously results in a perfect condition to be (irremediably) stuck in hole.

Accounting for this mechanism leads to a new strategy for stabilising the holes submitted to this lateral shift (no mud weight increase, etc... see detail in Ref. [44]). An exploration well drilled in 1992 with special care to avoid this deformation mechanism had good hole conditions down to a depth of 6909m, with a very low mud weight (1.15 to 1.20) to the great satisfaction of the geologists. It must be pointed out that following drilling incidents at an intermediate depth, a mud weight increase was revealed to be inefficient. Recommendations were given to lower the mud weight (amidst general scepticism) for the side-track which proved to be very efficient associated to adequate mud properties.

As a conclusion from these case histories, field cases coming from several major oil companies, and dealing mainly with drilling problems, were mentioned to illustrate the economic and safety impact of the Rock mechanics assessment of difficulties and problems in production. The following conclusions can be drawn from these examples :

- costs of direct and applied rock mechanics studies to a field case are currently in the range of half to few hundred thousand of US$. Amounts of money directly saved (see the domain of borehole stability) in case of conclusive studies are in the millions to tens of millions of US$

- applied to one field only, these studies have a "direct" return of more than 50 times the cost of the study, very often this ratio is greater than 100 or several hundred. The return is often greater because they apply to several fields

- Rock Mechanics evaluation is a very basic, fundamental, and mandatory step to identify the rock failure mechanism responsible for unexpected behaviour and induced problems. Then recommendations can be given for trajectories, drilling parameters (shapes of tools, bottomhole assemblies, trip procedures) and mud characteristics and properties

- once the mechanism is correctly identified, the remedy can be found quickly, adjusted, and be extraordinarily efficient. A huge amount of time can be saved, avoiding the successive trials always long and costly because needing successive corrections

- understanding the mechanisms allows us to extrapolate results to other situations and fields, which makes for an even higher return for the studies undertaken.

- **significant side (extra) benefits can be found in terms of safety** from a correct rock mechanics evaluation of field problems. They result also in terms of economics but are more difficult to appraise. It is difficult to quantify the savings arising from the avoidance of catastrophic consequences.

- extreme care must be taken in analysing the field incidents, observations and measurements : very often, indications on the rock failure (or unexpected behaviour, e.g. excessive deformation) mechanism are already present in general observations made by people in the field

- as regards only application to oil production, but similar conclusions could be drawn in Mining and Civil Engineering, in some specific areas such as wellbore stability, hydraulic fracturing, compaction and subsidence evaluation prediction, Rock Mechanics can bring determinant recommendations and remedies, once the basic mechanisms have been correctly identified. In other areas (drilling in shales), significant progress has been made resulting in considerable amounts of money saved.

4 ENCOURAGING RESULTS

Since the last symposium of the ISRM devoted to problems occurring at great depth in 1989, encouraging results were obtained :

- significant improvements have been made as regards the theoretical tools to describe the rock behaviour. Formulation of porothermoelastoplasticity, models derived from soil mechanics (for soft rocks at great depth), consideration of multiple rupture modes, now approached by bifurcation theory, identification and confirmation of failure by lateral shifts, ruptures induced by transient flow permits a much better separation and identification of various mechanisms of failure

- among them, progress is underway in shale behaviour. It must be recognised that these materials set together all the difficulties : anisotropy of deformability and strength, difficulty to evaluate the in situ state of stress, effect of low porosity and relevant problems in knowing the actual pore pressure and the effect of capillarity and desaturation, physico-chemical effects which are not fully understood

- progresses are also underway in monitoring systems, and namely borehole imagery tools in oil industry

- field cases are more often analysed in terms of basic mechanisms identification and then modelling : it is absolutely necessary to identify and separate the various mechanisms which can play a role. Consequences can be huge in terms of safety and economics as shown by field cases reported. This identification may result in adaptation or modification of design, construction or production methods or procedures, with many side benefits, some of them with a direct result on safety. When the mechanisms are not understood or not adequately evaluated (for instance transient state of stress and pore pressure during blasting, increase of water saturation for some porous material), reliable criteria cannot be proposed and adopted for design and construction.

5 REMAINING (PERSISTING) DIFFICULTIES FROM A ROCK MECHANICS PERSPECTIVE

When involved in design, project, construction, or expertise in underground excavations problems, Rock mechanics specialists are faced with various difficulties. Some of them are common to all the fields of industry, others not, but it is interesting to compare what can be found in some domains which can be useful for others.

(i) the knowledge of the **in situ state of stress** prior to excavation or production (oil) is still insufficient.

Around large ore deposits such as in South Africa, or large oil fields, in situ stress measurements or evaluation have been made by various methods and approaches. In these cases, the general state of stress style is known allowing an evaluation of conditions imposed to boreholes and excavations (there are certainly large variations at the local scale as in the faults vicinity or main geological accidents).

In most of the other cases, even the regional type of in situ state of stress ("vertical", "thrust", "horizontal" (or lateral) stress field according to vertical component being respectively major, intermediate, or minor principal stress, or even tilted states of principal stresses) is not sufficiently known. In many analyses, very large range of values of principal in situ stresses have to be considered, leaving uncertainties in the conclusions of studies and proposed remedies to problems.

Progresses has been made in some in situ stress measurements methods, such as minifrac tests procedure and interpretation. It can be hoped that more reliable data are going to be obtained in the future.

At a large scale, efforts such as the World Stress Map will probably allow to gather information at least at a regional scale. It may be questioned if Rock Mechanics specialists from mining, civil engineering and oil industry are sufficiently in contact with universities or research centres working on these subjects.

In Civil Engineering faced with crossing complex geological structures, the situation becomes worse due to certainly changing state of stress along the profile of the tunnels or galleries. The fact that the gallery is at atmospheric pressure involving similar modes of failure and the need for absolute safety involving high support and cladding must not hide the problems resulting from the poor knowledge of the in situ state of stress.

(ii) **basic rock failure mechanisms**

Even in apparently simple situations (drilled raised bored holes at atmospheric pressure, with no internal support), questions are still remaining about **the rupture features and occurrence at the wall** : true shear or curved column-style (extension) rupture, periodicity of ruptures lines, difference of behaviour in small and large diameter excavation. The example of various styles of spalling observed in gold mines in South Africa is self eloquent : in spite of attempts

made to better appraise the mode of rupture, the reason of periodic rupture lines by advanced theoretical approaches (bifurcation), there are not yet industrial tools to foresee conditions and frontiers between various ruptures styles. This question is also crucial in deep tunnelling.

Loading conditions - and varying loading conditions - are not sufficiently identified and recognised : the case of transient stress and pore pressure regime after blasting was mentioned, the example of oil wells where delayed ruptures (or excessive deformation) were sometimes attributed to creep of mysterious origin whereas thermal loading due to fluid circulation is very persuasive. Comparable situation is found with lateral shift induced by pore pressure diffusion in faults, giving delayed displacements (time of hydraulic diffusivity in faults) which are in no ways related to creep phenomena. An adequate rock failure mechanism identification is essential to solve these problems.

(iii) **combined mechanisms** : the question becomes even more critical when several mechanisms can interplay. An example is given by drilling oil wells in shales : rupture is the results of four or five possible loadings - namely in situ stress reconcentration, thermal stress (with pure thermo-mechanical and/or thermoporo-mechanical effects), internal pressure variation, interaction drilling fluid/rock - applied to anisotropic rock (deformability and strength). Interpretation of observations are impossible without being able to separate the effects of each parameter. This separation between possible mechanisms involved in a field case is probably the most difficult, fascinating but compulsory step to make in any situation.

(iv) **case of swelling rocks** : the origin of swelling is not yet perfectly understood, as well as the physical laws which govern it. The very different conditions that swelling rocks encounter in civil engineering (high deviatoric stresses and hydraulic gradients generated by the atmospheric pressure and the nil radial stress, absence of chemical effects due to the absence of drilling fluids) and in oil wells (less deviatoric stresses and gradients, but drilling fluid/rock possible interaction, all values of deviation and azimuths of the boreholes) suggests that closer co-operation should exist between field specialists working on these problems in both industries.

(v) **case of dynamic loading** : this type of loading - which in fact is a particular case of secondary loading due to stress redistribution - which is critical in deep mining is also present in some gas and oil fields. Progresses made in rupture location, observations, analyses by techniques derived from modern seismology have increased the knowledge of these failures. Nevertheless the prediction of this kind of failure still remains one of the most critical item specially for mines.

6 CONCLUSIONS, SUGGESTIONS

After several tens of years of participation in underground projects for mining, underground storage projects somewhat similar to civil engineering workings, and now with deep oil and gas wells, the author would like to take the liberty of making some suggestion or recommendations.

(i) literature devoted to rock mechanics problems encountered at great depth, sometimes considered as marginal in various industries (mining, civil engineering, oil production or exploration) is scattered among totally separate technical communities. It is somewhat difficult to be aware of progresses made in so different fields of application of rock mechanics. One of the first helpful tool could be to **have a global overview of the situation of deep underground works** (this one has to be considered as a very preliminary step), involving specialists coming from various industries to set up **a general bibliographic summary** (main places, general publications, main problems, main tools used, case histories, etc…). This key for access to « deep underground » literature is still missing (the short bibliography given here is only a very brief recall of some recent papers).

(ii) Technical recommendations concerning the above technical mentioned items are obvious, and progresses are underway in many research centres and universities to improve our knowledge of basic failure mechanisms, criteria, modelling, and new successes are certainly going to be obtained. We would only wish that maybe **more theses or research studies were undertaken in closer connection with field cases**.

(iii) The situation is less obvious for items needing in situ measurements (stress measurements, coring in oil industry), always time consuming and costly, and not always directly efficient. The example of shale coring - a shame for oil drillers - is particularly striking (why take a shale core, source of potential problems, possible bad recovery and uncertain use because of secondary breaking, implying 12 or 15 hours of rig (half a million US $). Is not it more profitable to (blindly) try another type of drilling fluid ?)

Although the impact of Rock Mechanics is now more valued in some mining or oil companies and for deep civil underground projects, it remains sometimes difficult to convince people in charge of operations to spend money for a coherent rock mechanics approach rather than a series of empirical trial and error exercises, extremely difficult to interpret, impossible to extrapolate to other situations, resulting in contradictory issues, source of endless and inconclusive discussions. Trying to save money on the direct costs for the execution and follow up of rock mechanics studies which cost several tens or hundred thousands of US$ implies accepting the loss of several MUS$ which is unacceptable.

In this purpose - and we will extend here the recommendations of Wagner to civil engineering and oil production -, it is strongly wished that rock mechanics specialists or the Rock Mechanics Community adopt **a strategy of disseminating the results** obtained and their impact in terms of costs and return. This must not be limited to the Rock Mechanics community, but **also directed to communities involved in the execution of workings and their corresponding technical groups**. Efforts have already been made in this direction, but must be developed and maintained. The author's experience, after trying to do so for several years for the attention of applied oil or gas technical communities showed that the effort must be on a long term basis, permanently maintained to observe the first results.

A systematic action of publicity of the results obtained would sometimes allow Rock mechanics specialists not to appear as odd people preoccupied with rather academical questions, asking for costly in situ measurements, the interpretation of which are questionable, and without profitable impacts to show. The first examples given here or by South African gold miners could be the type of approach to undertake more systematically.

Table 1 : COMPARED ROCK MECHANICS CONDITIONS and FEATURES OF WORKINGS AT GREAT DEPTH

Parameter:	MINING (permanent excavations)	CIVIL ENGINEERING (tunnels, adits, shafts, chambers)	OIL/GAS PRODUCTION (wells)
Internal pressure (in openings) :	Constant (atmospheric)	Constant (atmospheric) (exceptionally pressurised)	Highly variable (mud pressure, ΔP > 0 or < 0)
Temperature condition at the wall :	Cooling (constant)	Cooling (constant)	Highly variable (mud circulation, ΔT > 0 or < 0)
Hydraulic gradient around openings :	Constant (in principle)	Constant (in principle)	Variable (liquid or gas)
Support	Mechanical (Bolting, lining, etc...)	Mechanical	Hydraulic by mud pressure then Mechanical (casings)
In situ stress :	Homogeneous at large scale (scale of ore deposit but heterogeneous at small scale)	Possibly very variable along profile (scale of the rock mass to cross)	Homogeneous at large scale (scale of large reservoirs) / Heterogeneous in particular situations (foothills, salt domes)
Applied stress on excavation	Variable with progress of mining operations	Constant during life of working	Possibly variable with pressure depletion
Safety required :	Temporary	Definitive	Special requirement concerning control of fluid inflow

Table 2 : **FAILURE MECHANISMS OCCURRENCE, ORIGIN**

Failure/Deformation mechanism	MINING (shafts, passes, tunnels, chambers)	CIVIL ENGINEERING (tunnels, adits, shafts)	OIL/GAS PRODUCTION (wells)
Direct Matrix failure :			
In situ stress concentration around excavation	Spalling, strainburst (shear, extension)	Spalling, strainburst (shear, extension)	Various rupture modes (shear A, B, C, extension, traction)
Thermal stress	May have a minor effect	(Delayed failures ?)	Rupture due to heating, cooling
Hydraulic (fluid) gradient	Not applicable	Gouge material influx	Rupture due to transient gradient
Thermoporomechanical effect	Not applicable		Pressure gradient/thermal change
Drilling fluid/rock interaction	Not applicable		Drilling mud/shales interaction
Squeezing rocks	Possible in very localised zones	Delayed deformation/failure	Failures in evaporites
Swelling rocks	(rare)	Delayed deformation/failure	«
Geological joints, bedding	Unstable blocks	Unstable blocks + gouge material influx	Lateral shift by shear displacement
Interaction Matrix + Joints	Unstable blocks	Unstable blocks	Instability of fractured zones
Indirect (secondary) failure due to stress redistribution :			
Working progress and geometry change	d° + induced seismicity, rockburst	No	No
Operations (extraction/production)	d°	No	Effect of pressure decrease

[1]Cockerill I.D. « Safe, ultra-deep mining : current practice and future challenge » Proc. Colloquium Deep Level Mining - The Challenges, S. Afr. Inst. Min. Metall. 14 March 1996, 20p.

[2]Laubscher D.H. « Planning Mass Mining Operations » Comprehensive Rock Engineering Vol. 2 pp 547-583 J. Hudson Ed. Pergamon Press Publ. 1993

[3]Messines J.-P. & Guenot A. « Consolidation chimique des parois après forage à l'air » Proc. Int. Symp. ISRM/SPE Rock at Great Depth Pau 1989 Vol. 3, pp 1521-1533 Balkema Pub.

[4]Emmerich S.H. Report on rock pass problems in Anglo American Corporation Gold division Mines, Proc. Symp. on Orepasses and combustible materials underground, Ass. Min. Managers S. Afr., pp 83-111

[5]Gay N.C. « The stability of rock passes in deep mines » Proc. Symp. Rock passes and combustible materials underground, Ass. Min. Managers, S. Afr. pp. 128-163

[6]Gibbon J.A. « The repair of the main rock-pass systems at the Kloof Gold Mining Company Limited, Ass. Min. Managers, S. Afr. Papers & discussions 1976-1977 pp. 283-300

[7]Proc. Symp. on Rock Instability in Mine Shafts, S. Afr. National Group of ISRM Potshefstroom 1990

[8]More O'Ferral R.C. & Brinch G.H. « An approach to the design of tunnels for ultra-deep mining in the Klerksdorp district », Proc. Symp. Rock mechanics in the design of tunnels, S. Afr. Nat. Group of ISRM 1983 pp 61-65

[9]Jager A.J. Wojno L.N. & Henderson N.B. « New development in the design and support of tunnels under high stress » Proc. Int. Conf. Techn. Challenges in deep Level Mining S. Afr. Inst. Min. Metall. 1990 Vol.1 pp 1155-1172

[10]Ortlepp W.D. « Consideration in the design of support for deep hard rock « Proc. 5th Int. Tunnel Conference

[11]Stacey T.R. & Harte N.D. « Deep level raise boring - Prediction of rock problems » Int. Symp. ISRM/SPE Rock at great depth Pau France 1989 Vol. 2 pp 583-588 Balkema Pub.

[12]Panet M. « Quelques problèmes de Mécanique des Roches posés par le Tunnel du Mont-Blanc », Bull. Liaison Lab. Routiers des Ponts et Chaussées, 42 pp 115-145 1969

[13]Bois A.-P., Robert J. & Vuilleumier F. « Strainburst risk analysis in the Lötschberg base line tunnel. A case of history », Proc. 2nd North American Rock Mechanics Symposium 1996 pp 771-778

[14]Ortlepp W. D. & Stacey T. R. « Rockburst mechanisms in tunnels and shafts » Tunnelling & Underground Space technology 1994 Vol. 9, N° 1 pp 59-65

[15]Maury V. & Sauzay J.-M. « Borehole instability : Case history, Rock Mechanics Approach, and Results » SPE 16051 SPE/IADC 1986 Conf. New-Orleans pp 11-24

[16]Guenot A. « Contraintes et ruptures autour des forages pétroliers », Proc. 6th ISRM 1987 Congress, (G. Herger & S. Vangspaisal Ed. Balkema Pub. pp 109-118

[17]Maury V. « An overview of Tunnel, Underground Excavations and Borehole Collapse mechanisms » in Comprehensive Rock Engineering, Vol. 4 Ch. 14 pp 369-412 J. Hudson Ed. Pergamon Press

[18]Maury V.& Idelovici J.-L. « Safe drilling of HP/HT wells, The role of the thermal regime in loss and gain Phenomenon » Pap. SPE 29428 presented at the SPE/IADC Conf. Amsterdam Proc. pp 819-829

[19]Fabre D., Goy L., Menard G. & Burlet D. « Températures et contraintes dans les massifs rocheux : cas du projet de Tunnel Maurienne-Ambin » Tunnels et ouvrages souterrains, 1996 **134** pp 85-92

[20]Maury V., « Effondrements spontanés - Synthèse d'observations et possibilité de mécanisme initiateur par mise en charge hydraulique », Revue de l'Industrie Minérale Oct. 1979

[21]Panet M. « Le calcul des tunnels par la méthode convergence-confinement » Presses de l'Ecole des Ponts et Chaussées, Paris 1995

[22]Barla G. « Squeezing rocks in tunnels » ISRM News 1994 **2**, 2/3 pp 44-49

[23]Einstein H. ISRM Commission on Swelling Rocks, « Comments and Recommendations on design and analysis procedures for structures in argillaceous rocks » Int. Journ. of Rock Mechanics & Mining Sc. & Geomech. Abstr. 1994 **31**, 5 pp 535-546

[24]Aydan O., Akagi T. & Kawamoto T. (1993) « The squeezing potential of rock around tunnels : Theory and prediction » Rock Mechanics and Rock Engineering, **26**, 2 pp 137-163

[25]Sato J., Dong J.-J. & Aydan O. & Akagi T. (1995) « Prediction of time-dependent behaviour of a tunnel in squeezing rocks » 4th Field measurement in Geomechanics Symposium pp. 47-54

[26]Pan Y.-W., Dong J.-J. 1991 a & b « Time dependent tunnel convergence Part I Formulation of the model, Part II Advance rate and tunnel-support interaction » Intern. J. of Rock Mech. & Mining Sc. & Geomech. Abstr. 28, 6 pp 469-488

[27]Sahores J. « Contribution à l'étude des phénomènes mécaniques accompagnant l'hydratation de l'anhydrite » Thèse Univ. Toulouse (France), Rev. Matériaux de construction Pub. Techn. **126** (1962)

[28]Madsen N.T. & Nuesch R. 1991 « The swelling of clay-sulfate rocks » 7th ISRM Congress pp 285-288

[29]Nüesch R., Madsen F.T. & Steiner W. (1991) « Long term swelling of anhydritic rocks : Mineralogical and microstructural evaluation » 8th ISRM Congress pp 133-138

[30]Kovari K., Amstad Ch. & Anagnostou G. 1988 « Design/construction methods - tunnelling in swelling rocks » 29th US Symp. Rock Mechanics pp 17-32

[31]Ortlepp W.D. « High ground displacement velocities associated with Rockburst damage » Proc. 3rd Int. Symp. On Rockbursts and Seismicity in Mines, (1993) Kingston, Ontario pp 101-106 Balkema Pub.

[32]Stacey T.R. & Ortlepp W.D. « Rockburst mechanisms and tunnel support in rockburst conditions » Proc. Int. Conf. geomechanics 93 (1993) pp 39-46 Balkema Pub.

[33]Stacey T.R., Ortlepp W.D. & Kirsten H.A.D. « Energy-absorbing capacity of reinforced shotcrete, with reference to the containment of rockbursts damages » Journ. South Afr. Inst. Min. Metall. Vol. 95 pp 137-140

[34]Ortlepp W.D. & Stacey T.R. « Performance of rock containment support such as wire mesh under simulated rockburst loading » Research report (at press 1996)

[35]Maury V., Grasso J.-R. & Wittlinger G. « Monitoring of subsidence and induced seismicity in the Lacq Gas Field (France) : the consequences on gas production and field operation », Engineering Geology 1992 **32** , pp 123- 135

[36]Fourmaintraux D., Grasso J.-R. , Bard P.-H. & Koller M. « Use of continuous seismic monitoring for hazard assessment of seismicity associated with hydrocarbons reservoir and triggered by production » Proc. 8th ISRM Congress 1995 Tokyo Fujii Ed. pp 915-922

[37]Smirnova M.N., Shebalin N.V., Navitskaïa N.A. & Miatosckin V.J. « Earthquake swarm caused by exploitation of an oil field » IUGG **16** Gen. Assembly Grenoble 1975

[38] South African Nat. Group of the ISRM (SANGORM) Proc. of the Sangorm Symp. Welkom Oct. 1991 « Impact of Rock Engineering on Mining and Tunnelling Economics »

[39]Maury V. « Rock Mechanics over the last ten years in Oil and Gas Production : Safety and Economy » Proc. 14th World Petroleum Congress, 1994 J. Wiley & Sons Pub. pp 203-213

[40]Maury V., & Guenot A., "Practical advantages of mud cooling systems" SPE Drilling & Completion March 1995 pp 42-48

[41]Ewy R.T., Ross G.D., Gast M.R., Steiger R.P., "North Sea Case histories of wellbore stability prediction for successfull high-angle Nelson Field wells", Paper IADC/SPE 27495 Dallas Feb. 1994

[42]Wong S.W., Kenter C.J., Schokkenbroeke H, de Bordes P., "Optimizing shale drilling in the Northern North Sea: borehole stability considerations", Offshore Eur. Conf. Aberdeen Sept; 1993, pp.493-505

[43]Maury V., Sauzay J.-M.,"Rupture de puits provoquée par glissement sur failles: Cas vécu, mécanisme, remèdes, conséquences", Proc. Int. Symp. ISRM/SPE Rock at great depth, Pau 1989, Vol. 2, pp.871-882, Balkema Pub.

[44]Maury V. & Zurdo Ch., "Drilling-Induced Lateral Shifts along Pre-existing fractures : A common cause of Drilling problems » SPE Drilling & Completion, March 1996 pp 17-23

Eurock '96, Barla (ed.) © 2000 Balkema, Rotterdam, ISBN 90 5809 843 6

Rock mechanics approaches for understanding flow and transport pathways

W.S. Dershowitz

Golder Associates Incorporated, Seattle, Wash., USA

ABSTRACT: Rock mechanics approaches are frequently distinguished by their emphasis on the effects of geological structure. This is particularly important in understanding flow and transport pathways through fractured rock. This paper presents recent field measurements which highlight the importance of geological structure in defining flow and transport pathways, and presents the theoretical development of "pathways analysis" approaches to these problems.

1 OVERVIEW

In a conventional porous medium, groundwater flow and transport can be understood in terms of a flow net controlled by the combination of the hydraulic conductivity field and the in situ head field (Bear, 1972). Experience over the past 20 years has demonstrated that this approach is inadequate for many fractured rock masses, in which flow and transport is structurally controlled. For illustration, this paper presents a few examples of the unique hydraulic behavior of structurally controlled rock masses. For these rock masses, flow and transport pathways follow the geometric connections of the discrete features, rather than the smooth contours of the idealized porous medium "flow net" (Figure 1). As a result, the rock mechanics emphasis on discontinuities for geomechanical analysis also plays a key role in environmental applications. This is seen most clearly in the discrete feature network (DFN) approach (Dershowitz, 1984, Robinson, 1984).

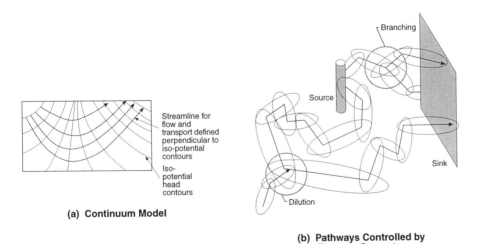

Streamline for flow and transport defined perpendicular to iso-potential contours

Iso-potential head contours

(a) Continuum Model

Branching

Source

Sink

Dilution

(b) Pathways Controlled by Fracture Geometry

Figure 1. Continuum streamline and fractured rock transport pathways.

The DFN approach has recently been extended to directly address the geometry and properties of flow and transport pathways based on rock mechanics analysis, rather than focusing on conventional flow and transport solutions through fractured rock (Dershowitz and Roberds, 1990). Following the illustrative examples of discrete feature controlled behavior, this paper develops the theoretical basis of the pathways analysis approach, and provides example applications.

2 FLELD OBSERVATIONS

In considering the rock mechanics aspects of environmental problems, it is interesting to note that abandoned mines are increasingly being used as landfills, while regulations are increasingly focusing on identification and monitoring of potential transport pathways. For example, in Canada, where a number of landfills are currently being sited or studied for siting in abandoned open pit mines, pathway identification and monitorability has become a major issue (Ministry of Energy and Environment, 1996). The monitoring requirement has made it necessary to use gradient control to ensure that all landfill leachate can be collected within the mine. This represents a major technical issue for agencies siting landfills, and one which can only be addressed by explicitly analyzing the rock mechanics issues.

Structurally controlled flow and transport pathways have been observed in a wide variety of geological media. The controlling structures include fractures, stratigraphic bedding, and faults, karsts and other solution features, and sedimentary structures such as paleo-channels. One consequence of structural control of flow and transport is that understanding of the rock matrix itself becomes less important relative to an understanding of the discrete features. This concept is a comer stone of rock mechanics.

2.1 Discrete flow: Key Lake Mine

The Key Lake Mine in Northern Saskatchewan had been using porous media flow concepts as the basis of their pit stability and dewatering control scheme during excavation of the loose overburden and sandstone units. However, during excavation and mining of uranium into the precambrian basement rock, it was discovered that the most effective way to provide dewatering and ensure pit slope stability is through horizontal wells intersecting the main

conductive structures. At the Key Lake Mine this was primarily the unconformity between the Precambrian basement rock and the overlying sandstone. Hydro-fracturing of this conductive structure during 1996 through a vertical borehole offset some 50 meters from the high wall resulted in increase inflows via the horizontal wells such that other inflows to the open pit are insignificant by comparison. Figure 2 illustrates the magnitude of the flows observed from the structural unconformity, as compared to the flow to standard horizontal wells.

courtesy of Scott Donald, Golder Associates Ltd.

Figure 2. Flow from horizontal well intersecting structural unconformity.

2.2 Fracture transport pathways: Ramat Hovav Site, Israel

The Ramat Hovav facility south of Beer Sheva is the major chemical waste disposal in Israel. Located on low permeability Dolomite formations, conventional analyses indicated that temporary chemical waste overflow storage could be safely provided in unlined ponds, depending on the generally low permeability of the bedrock. The concept of porous medium analysis was supported by the high fracture frequency (on the order of one per meter), and low average effective rock mass permeability.

Figure 3 shows the fracture which provides a discrete pathway which transported waste over a distance of approximately 200 meters from the pond to the neighboring seasonal stream or "wadi" where the fracture is shown discharging. The photograph in Figure 3 shows the chemical rind on the fractures providing the pathway. This pathway is formed by a single, discrete 200 m scale fracture, together with a small network of intersecting fractures. Deposits can be clearly seen on the walls of the fracture and on the rock immediately adjacent to the fracture. These deposits are indicative of solute transport through the discrete fracture. The porous media analysis was correct in predicting that the rock mass would provide an adequate barrier to movement of a conventional plume. However, a discrete feature analysis based on conventional rock mechanics and structural geological interpretations would have indicated the probability of occurrence of discrete pathways through fractures and fracture networks.

Figure 3. Chemical waste transport pathway in 100 m scale fracture.

2.3 Karst/Wormhole transport pathways: Oak Ridge, Tennessee

A similar phenomena has been observed at the Oak Ridge, Tennessee site, where discrete feature flow occurs at scales from meters to kilometers. Tracer testing in the Bear Creek area during 1986 demonstrated that tracer followed a linear path defined by a combination of bedding planes and fractures, rather than following a porous medium plume (Pearson and Lozier, 1986). While analysis of the geologic structure predicted the tracer transport path, calibrated continuum models had great difficulty. Pressure measurements made in the Bear Creek valley at the same time indicate the occurrence of isolated pressures at depth which are more than 50 meters less than the heads in the surrounding rocks. This is consistent with the occurrence of solution features providing hydraulic connections over distances of hundreds of meters at least.

More recently, a major tracer testing program has identified a highly radioactive plume which has traveled several hundred meters, confined exclusively to a fractured interbed (Figure 4). The key to isolation and remediation of this plume is the application of rock mechanics approaches to characterize the geometry of the structural features which control the transport pathway geometry for all of the pathways and related pathway branches (Ketelle, 1996).

2.4 Compartmentalization: Kamaishi Mine, Japan

One of the more intriguing environmental aspects of fractured rock is the occurrence of isolated "compartments", which restrict flow and contaminant transport, despite high permeability (Doe et al., 1996). The Kamaishi site is a former iron mine in north-eastern Japan, on the Pacific coast. It is currently operated as a research laboratory for the Japanese radioactive waste disposal authority, PNC.

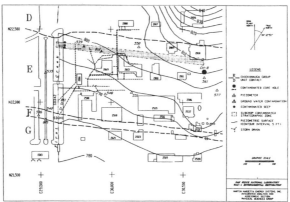

(after Ketelle 1996)

Figure 4. Transport pathway at ORNL.

During 1994, anomalously high head values were observed close to experimental drifts. A series of boreholes were installed and tested to characterize the flow regime within a 100 meter scale rock block. A continuous monitoring program was installed to measure cross-hole response during installation of each successive borehole (Doe, et al., 1996).

Figure 5 presents a plot of the heads recorded in a 100 meter block of rock adjacent to the mine. Six distinct head zones can be distinguished, with heads ranging from 100 meters to 10 meters, without direct correlation of the distance to the drift, where head is referenced to O meters. Analysis of transient response during drilling, and subsequent cross-hole hydraulic testing (Doe, et al., 1996) demonstrates that the fractures within each "compartment" behave in unison, and that transients in one compartment are considerably reduced to even not measurable in other compartments.

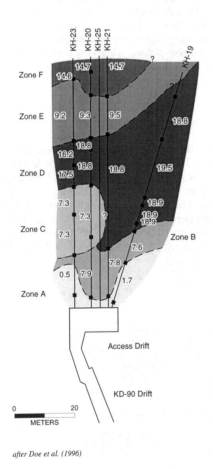

after Doe et al. (1996)

Figure 5. Compartmentalization at Kamaishi.

From the perspective of environmental rock mechanics, this has both positive and negative implications. Contamination which has entered into a compartment will not leave that compartment. As a result, large scale contamination will be more limited in a compartmentalized rock mass than in either conventional continuum system or well connected fractured rocks. On the negative side, monitoring and remediation of such a site will depend on identification of the effected compartments, and investigation of uncontaminated compartments could lead to unrealistically optimistic assessments.

Doe et al. (1996) has hypothesized that hydraulic compartments can occur in fracture networks as a result of fractal fracture location processes, which can produce isolated fracture networks (Figure 6). Alternatively, these clusters may be the result of fracture mineralization processes which clog specific hydraulic connections where changes in groundwater or rock chemistry occur, causing compartmentalization. Or, compartmentalization can occur as a result of clay gouge related to shear movements on faults.

Analysis of the structurally controlled geometry of transport pathways in compartmentalized rock masses requires an understanding of both the geometry of the individual fractures, and the geometry of the connections between the compartments, and the hydraulic barriers separating the compartments. This requires rock mechanics analysis of the discrete features which make up the pathways.

2.5 Compartmentalization: Yates Field

While the compartmentalization at Kamaishi occurred in a granitic rock with a high fracture intensity and negligible matrix permeability, compartmentalization at the Yates Field in West Texas occurs in a dolomite/limestone reservoir with a relatively low fracture intensity and a permeable matrix. Apparent flow compartmentalization at Yates appears when there is a significant imbalance in fluid withdrawal and injection. Variation in the degree of fracture connection and solution enhancement control this type of flow compartmentalization, since minimal flow occurs across the permeable matrix.

Improved production has been achieved at Yates by recognizing the need to engineer production in terms of (a) the hydraulic compartments, (b) the fracture-controlled pathways which provide production to the wells, and (c) the use of thermally-assisted gravity segregation (TAGS) to stimulate the matrix adjacent to key pathways (Wadleigh, 1995).

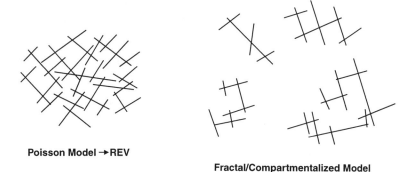

Poisson Model → REV

Fractal/Compartmentalized Model

Figure 6. Compartmentalization of fractal fracture patterns.

2.6 *Heterogeneous connectivity: Wairakei geothermal field*

In evaluating the performance of the Wairakei, New Zealand and other fractured rock geothermal fields, Horne and Rodriguez (1983) noted that tracers could arrive at locations kilometers away within hours, while within the same field other locations show slow or no tracer recovery. This heterogeneous connectivity is the other side of compartmentalized behavior. In many fractured rocks, single discrete features and fracture networks can produce isolated hydraulic and solute transport pathways, unaffected by the surrounding rock mass. This is very important for understanding environmental as well as geothermal pathways.

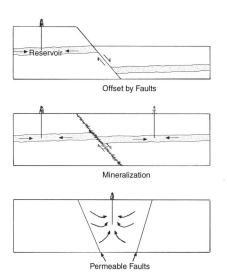

Figure 7. Oil reservoir compartmentalization.

The extremely rapid solute transport observed in geothermal reservoir discrete pathways has very serious implications for environmental application. This is not to imply that rock cannot provide good isolation for hazardous and radioactive wastes, but rather that the containment properties of rock need to be evaluated with explicit consideration of the potential for fast pathways defined by single features or simple networks of interconnected discrete features.

2.7 *Heterogeneous connectivity: Äspö Mine, Sweden*

The final example of heterogeneous connectivity is from the Äspö (Sweden) Hard Rock research laboratory operated by SKB, the Swedish radioactive waste disposal authority. In late 1991 and early 1992, SKB constructed an elevator shaft to their underground research laboratory (Forsmark and Rhen, 1993). SKB monitored the pressure response using a network of over 200 pressure measurement instruments installed on Äspö island and the adjacent areas. Figure 8 illustrates the three distinct types of response were observed (Uchida et al., 1996b):

- Gradual Response: Typical monitoring intervals displayed a gradual, "Theis" type response, responding to the construction of the shaft over a period of several months.
- No Response: A number of intervals, particularly those intervals monitored under the Baltic, showed no response, either because the intervals were well connected to a hydraulic boundary or because they were not hydraulically connected to the shaft.

- Step Response: About 20 monitoring intervals responded as a "step function" to the construction of the shaft.

The spatial distribution of the different types of response is illustrated in Figures 9 and 10. 3-D analysis of the spatial distribution of "Step" responses indicates that these occur on a single discrete plane (Figure 11). Note that the rapid, hydraulically isolated step response occurs within this feature at scales of over 300 meters. This behavior is consistent with a single fracture zone with high diffusivity, but poor connections with intersecting fractures and fracture zones. This poor connection could be due to, for example, clay gouge in the structure of the fracture zone.

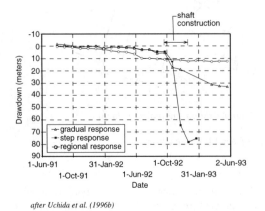

after Uchida et al. (1996b)

Figure 8. Responses to shaft construction at Äspö.

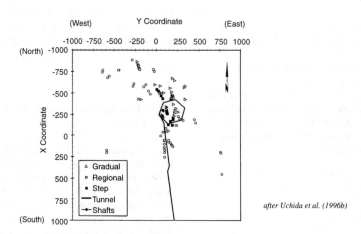

after Uchida et al. (1996b)

Figure 9. Plan view of hydraulic responses to shaft construction.

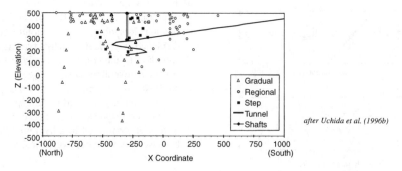

after Uchida et al. (1996b)

Figure 10. North-South cross section of hydraulic responses to shaft construction.

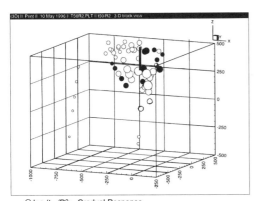

○ log (t_{50}/R²) - Gradual Response
● log (t_{50}/R²) - Step Response
   Size of symbols scales linearly to log (t_{50}/R²)
   Vertical shafts and tunnels are shown

after Uchida et al. (1996b)

Figure 11. 3-D view of "Step" hydraulic responses.

The existence of this zone is significant, since it was not detected by hydraulic investigations during the period 1987-1995. The behavior of this zone appears to be unique, although its orientation is consistent with a NNW trending set of sub-parallel fracture zones (Figure 12). Responses in other NNW trending fracture zones generally show Theis responses, consistent with zones which are well interconnected with the other fracture zones, as would be expected from the density of fracture zones shown in Figure 12.

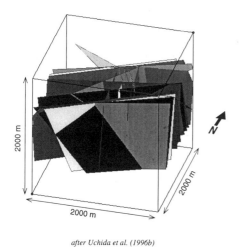

after Uchida et al. (1996b)

Figure 12. Äspö Fracture Zones.

The response in the zone characterized by step response to shaft construction again indicates a need to use rock mechanics approaches to define the discrete features at the site. However, the lack of direct characterization of the zone except for its distinctive hydraulic response to shaft construction indicates the care which needs to be taken in interpreting pathways defined by structural features. Although this feature is intersected by 6 significant fracture zones on Äspö island, none of these provide a hydraulic connection sufficient to dampen the step response to shaft construction.

3 PATHWAYS ANALYSIS MATHEMATICAL DEVELOPMENT

Realistic analysis of the fractured rock behaviors described in the previous section is essential to proper engineering for environmental applications. This section describes the development of a simplified engineering method, "Pathways Analysis", which is designed to provide approximate solutions for environmental problems in fractured rock.

Pathways analysis pre-supposes that a realistic, discrete fracture network (DFN) model has been developed, based on conventional fracture statistics and DFN hydrological interpretations. For a description of the procedures for implementation of DFN models, see, for example, Dershowitz et al., 1995.

A DFN model consists of a 3-D volume containing a combination of deterministic, conditioned, and stochastic 2-D discrete features representing fractures, faults, karsts, and other discrete features (Figure 13). These models are based on definition of one or more sets, where each set is defined by a spatial location process, and distributions for orientation, size, shape, and hydrogeological and geomechanical properties. The flow and transport in a DFN model can, of course be solved explicitly using a finite element flow and transport solver, such as NAPSAC (Roberts, 1992), or MAFIC (Miller et al., 1995). However, the procedures and computational requirements of finite element flow and transport modeling are sometimes beyond the scope of practical applications. For these applications, the practical alternatives are sometimes reduced to the use of continuum models, which neglect the typical rock mass behaviors described in Section 2.0, or approximate methods. Pathways analysis provides an attractive approximate alternative for many applications.

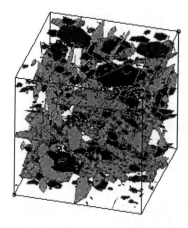

Figure 13. 3-D discrete feature network model.

The geometry of transport pathways in fractured rock is controlled by the geometry of discrete fractures. The approach proposed for the derivation of the geometry and properties of the key or effective hydraulic connections within the rock mass by pathways analysis is illustrated in Figure 14. First, a DFN model is used to generate realistic fracture geometries. The interconnected network of 2-D fractures is then converted to a topologically equivalent network of 1-D pipes. The head field within the network of 1-D pipes can be defined using either a linear circuit approximation, a conventional finite element or LaPlace Transform Galerkin solution. Pathways analysis can then use a graph-theory search to directly define the environmental implications of the fractured network in terms of transport pathways.

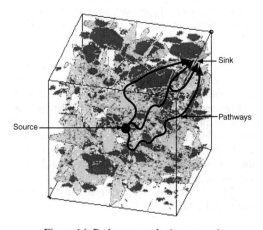

Figure 14. Pathways analysis approach.

3.1 Pipe approximation for fracture network topology

For pathways analysis to be easier and faster than DFN flow and transport modeling, fracture networks need to be simplified as a three dimensional network of 1-D pipes. This requires a significant simplification of the topology of the fracture network.

The algorithm for generating networks of pipes from the plate fracture network is as follows:

1. Calculate intersections between every fracture in the DFN to develop a linked list of fracture connections. This reduces the topology of the three dimensional fracture pattern to a series of nodes and links, where nodes represent the intersections between fractures, and links represent the fractures between each pair of nodes. Nodes are defined at the midpoint of each fracture intersection fibre, and also at any point where pipes intersect.

2. Within each fracture, define pipes between every pair of nodes.

3. Calculate the following properties for each pipe:
   - Pipe length L_m: the distance between the centers of the traces formed by the intersections of fractures F_{m-i} and F_{m+i} with fracture F_m
   - Pipe flow width W_m: the average trace length for the two traces
   - Pipe transport aperture a_m: the transport aperture from the FracWorks simulation
   - Pipe flow surface area A_m: the surface area available for flow (W_m x L_m)
   - Pipe flow conductance C_m: the conductance for the pipe (W_m x T_m).

Additional properties can also be calculated such as the volume of rock within a distance x of the pipe, and the area of the fracture plane which is accessible to the fluids in the pipe through diffusion.

3.2 Approximate head solutions

In the simplest applications, pathways analysis can carry out searches based on pipe conductance or travel time, without requiring a head solution. However, for many applications, it is useful to obtain at least an approximate head solution before carrying out the search. Alternative approximate head field solutions include:
- series flow approximation
- finite element solution, and
- LaPlace Transform Galerkin solution (Sudicky and McLaren, 1989).

The series flow approximation calculates the heads

at each node along each pathway by assuming linear, series flow between the sources and sinks at each end of the pathway.

For the series flow solution, the effective conductance C_k for a pathway k from a source to a sink is,

$$C_k = \frac{L_k}{\sum_m \left(\frac{L_m}{C_m}\right)}$$

where the pathway length L_k is calculated as the sum of the lengths of the pipes making up the path,

$$L_k = \Sigma L_m$$

Flow through the pathway is therefore:

$$Q_k = C_k \, (h_{source} - h_{sink})/L_k.$$

Since the flux through each part of a linear pathway must be constant and equal to the flow through the pathway, the head drop Δh_m across each pipe m can be calculated from:

$$Q_m = C_m \Delta h_m / L_m$$
$$\Delta h_m = Q_m L_m / C_m$$

Once the head drop across each pipe is known, the heads h_m at each node m can be found from the heads at the up-gradient node h_{m+1} as:

$$h_m = h_{m-1} + dh_{m-1}$$

Note that this approach does not ensure that the head at each node will be the same for flow through the node on different pathways or in different directions. As a result, this method for calculating heads must be considered as a very approximate approach.

The finite element pipe network and LaPlace Transform Galerkin head solutions both provide exact solutions for the head within the pipe network, at the expense of significantly greater computational effort than the series flow solution. In addition, the flow solution is still approximate when compared to the head solution for the 3-D fracture network.

3.3 Pathways search algorithms

The key to the pathways approach is the use of geometric information to identify the interconnected discrete features which provide the key transport pathways. This can be implemented using a variety of algorithms from computational geometry. Sedgewick (1988) presents a variety of algorithms, including graph theory "depth first" and "breadth first" pathway searches which have application to the identification of environmental pathways. This sections describes these two major types of search algorithms, and their advantages. First, however, we need to address the definition of search criteria.

Pathways analysis searches are based on specification of "sources" and "sinks". The sources can be, for example, contaminated sites or injection locations, while the sinks can be, for example, discharge locations, compliance boundaries, or wells. For the series flow solution, pathways need to be identified before calculating heads. As a result, pathways are based purely on the geometric properties. For head solutions which calculate approximate heads before definition of pathways, the search criteria can directly incorporate flux and similar criteria as search criteria.

Possible search criteria include, for example:

- Maximum effective conductance (based on the effective series-flow pathway conductance).
- Minimum travel time (based on the sum of the travel times through each of the pipes which make up the pathway).
- Maximum flux, defined as the flux at the source node, or
- Minimum dilution, where dilution is calculated as the ratio of the flux through the pipe at the sink node divided by the flux at the source node.

The depth first search is based on Sedgewick (1988). The search algorithm identifies a set of non-reentrant paths, including the "highest priority" path, based on the criteria established by the user. The algorithm does not identify all combinations of possible paths.

The depth-first, priority order search is as follows:

- All nodes connected to the source are identified, and the node representing the fracture with the highest value of the user specified "priority" ranking is selected and marked.
- All of the unmarked nodes connected to the selected node are identified.
- The process is repeated until either a dead end or the sink is reached. When the sink is reached, all of the links between node pairs which makeup the pathway are marked "visited". Each node can only be "visited" once in a pathway.
- As each pathway is identified, the pathway properties are calculated using the equations below.
- Significant branches of each pathway are then identified by applying the priority criteria to the branches of the identified pathway, working ba-

ckwards from the sink.

- The process is repeated, moving backward from the sink as the search progresses.

While the depth first search produces a "tree" structure of branching overlapping pathways, the breadth-first search produces a minimum set of pathways which completely avoid overlap. Once a pipe is used in a path, it is removed from consideration from inclusion in any other path. As a result, the breadth first algorithm produces a minimum set of pathways.

In the breadth-first search, all searches begin at nodes intersecting the source (Sedgewick, 1988). The search algorithm identifies path in the order based on the "priority" order criteria established by the user. The algorithm does not identify all combinations of possible paths.

The breadth-first, priority search algorithm is as follows:

- All nodes connected to the source are identified, and the node representing the fracture with the value of the user-specified priority ranking is selected and marked.
- All of the unmarked nodes connected to the selected node are identified.
- The process is repeated until either a dead end or the sink is reached. When the sink is reached, all of the links between node pairs which makeup the pathway are marked "visited", and are excluded from future searches. Unmarked nodes from those connected to the path are then selected, and the process is repeated.
- As each pathway is identified, the pathway properties are calculated using the equations below.

3.4 *Pathway properties*

To identify pathways based on criteria such as those listed above, and to describe the properties of pathways once they are identified, requires the definition of "equivalent pathway properties". These properties can the be used with 1-D transport solutions to obtain approximate tracer transport solutions, based on the effective properties. Examples of effective pathway properties include:

- Pathway conductance C_m, the flux through the pathway per unit of gradient (m^3/s).
- Pathway area A_m, the effective area for retardation/sorption or diffusive processes along the pathway. Due to turbulent and diffusive mixing along pathways, this value is between the total area of all fractures touched by the pathway and the total surface areas of the transport channels

formed within the pathways. This can be expressed as an area per pathway (m^2), or can be normalized by the rock volume (m^2/m^3).

- Hydraulic radius r_m, the radius for a tube which would have conductance C_m according to, for example, the tube flow equation or the "cubic law". For the tube flow equation in a tube or width W_m with transmissivity T_m,

$$C_m = T_m \, W_m$$

$$r_m = \left(\frac{8\mu C_m}{\rho g} \right)^{1/4}$$

- Advective travel time t_m through pathway, which might be calculated as the sum of the travel times through the individual fractures which make up the pathway.
- Penetration depth B_m, the depth to which the rock is available for processes such as matrix diffusion, consistent with the advective travel time, t_m and surface area, A_m. This might be calculated by geometric analysis of the rock blocks defined by the fracture pattern.

4 PATHWAYS ANALYSIS ENVIRONMENTAL APPLICATIONS

Pathways analysis is useful in a wide range of environmental rock mechanics applications. This section describes six typical applications.

4.1 *Well-head protection*

Well-head protection is concerned with the probability of aquifer contamination from surface water infiltration and groundwater mixing in inadequately isolated wells. Groundwater contamination from wells is directly controlled by the interconnections of conductive fractures with the well. Well-head protection in fractured rock therefore needs to be concerned more with the geometry of the fracture network than with, for example, the specific capacity of the well. If all of the specific capacity is related to a single fracture, then the well can be controlled by isolating the intersection with that fracture, while a well with many significant connections might require multiple installations. Rock mechanics can therefore provide significant benefit for well-head protection, through analysis and prediction of the nature of well interconnections, and the extent and properties of networks connected to wells. These networks can be directly analyzed according to pathways analysis as described in Section 3.

4.2 Contaminant plume pump-and-treat

Typical pump-and-treat installations are based on the assumptions that the pumping well is accessing and cleaning a cylindrical volume of rock surrounding the well. In fractured rocks, however, the effectiveness of pump-and-treat depends on (a) the ability of the defined well array to access the entire contaminated fracture network, and (b) the ability of the fracture networks connected to the pumping well to remove contamination from adjoining rock blocks. This is a classic connectivity problem, well suited for pathways analysis.

The analysis of pump-and-treat installations using pathways analysis can be carried out as follows:
1. Define the geometry and hydraulic properties of fractures in the rock to be treated.
2. Generate synthetic geologic realizations based on the site fracture geometry as a DFN model.
3. Simulate the alternative pump-and-treat process designs in the rock using DFN flow and transport modeling, or with pathways analysis.
4. Define the appropriate pump-and-treat process.

This approach can be refined and repeated as additional information is collected during borehole installation and test pumping.

4.3 Tracer test design and interpretation

Tracer test design in fractured rock is complicated by both the complexity of the fracture interconnections, and by the complex head fields which result from fracture network connections. Tracer test design in fractured rock must therefore explicitly consider these connections and the affected gradients.

For example, a recent tracer test carried out at Äspö, Sweden (Winberg et al., 1996) was designed to study transport under the minimum possible gradient in a single fracture. However, fracture studies (Dershowitz and Busse, 1996) indicated a high degree of fracture interconnection and the presence of complex gradients in the tested fracture, with a significant possibility that tracer would not migrate from the injection borehole to the pumping borehole (Figure 15). Meanwhile, confined aquifer continuum models uniformly predict 100% ultimate tracer recovery. The experiment, however, found that tracer was recovered to the extraction well from only three of the four injection boreholes (Figure 16). The remaining tracer was drawn by gradients into interconnecting fractures (Winberg, 1996).

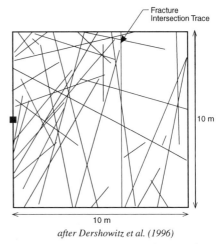

after Dershowitz et al. (1996)

Figure 15. Fracture intersections with "2D feature" for tracer experiment.

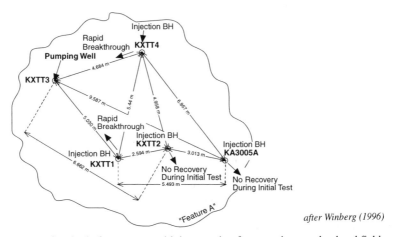

after Winberg (1996)

Figure 16. Tracer recovery for single fracture test with intersecting fractures in complex head field.

Tracer test design in fractured rock thus needs to be carried out with full consideration of fracture network interconnections, and their effect on head fields. While strong gradients can be used in tracer tests to force recovery, the consequence is that the resulting tracer pathways may not be representative of potentially more complex pathways under natural gradients. Tracer test design and interpretation should therefore consider the geometry of the transport pathways, both in establishing the geometry of the experiment, and in calculating the head field under which the experiment will be carried out. Depending on the application, it may be desirable to minimize induced gradients to obtain additional information regarding in situ conditions.

4.4 Evaluation of flow and transport barriers

Hydraulic compartmentalization and flow barriers such as those observed at Kamaishi and elsewhere can play a significant role in environmental rock mechanics, isolating wastes and preventing the spread of contamination. As a result, it is essential to include exploration for flow barriers in the characterization of contaminated sites, design of remediation, and analysis of waste storage facilities. Natural hydraulic and transport barriers are characterized by large local gradients (Doe et al., 1996), rapidly changing groundwater chemistry (Mazor, 1990), and distinct hydrograph transient signatures (Doe, et at., 1966).

Thus, while major investments are regularly made in artificial flow and transport barriers such as grout curtains and cutoff walls, natural flow barriers are frequently overlooked in conventional hydrologic analysis. This is in part because exploration programs are designed to characterize homogenous, continuum aquifers, rather than discrete flow systems. It is frequently useful to go back over site characterization results, looking specifically for evidence of natural flow and transport barriers which can be taken advantage of in environmental assessment.

4.5 Transport pathway identification

Environmental analyses frequently focus on identifying contaminant transport pathways (Dershowitz and Roberds, 1990). However, many of these analyses mistakenly focus on characterizing continuum "plumes", when the transport path is actually defined by discrete features in fractured rock. Transport pathways in fractured rock can be distinguished as "advective" or "diffusive" (Figure 17). The ad-

vective pathways are generally only the most conductive discrete features, and provide the fastest and most extensive pathways. In hard rock, the diffusive pathways are defined by the lower conductance fractures and the surrounding rock mass (if it has sufficient porosity). Pathways analysis can be used to identify the key advective pathways, and to characterize the diffusive pathways. If the rock contains well characterized, significant discrete features such as fracture zones and faults, pathways analysis may provide deterministic identification of the key pathways. Generally, however, discrete features are only known in a statistical sense, such that the pathways analysis must be carried out probabilistically. The pathways analysis can be carried out on multiple realizations of possible fracture geometries to define the pathway geometry, effective transport properties, and expected discharge locations.

4.6 Design of dewatering

Groundwater depressurization is a key aspect of slope stability, particularly for open pit mines. However, while rock mechanics slope stability generally is based on discrete fracture concepts, slope dewatering rarely uses DFN procedures directly. This may in part be due to the perception that DFN analysis is expensive and time consuming. However, pathways analysis provides an opportunity for rapid, approximate DFN analysis while maintaining many of the more important features, as was shown in the Key Lake example presented above.

Pathways analysis can provide the following tools for dewatering design:

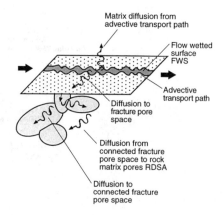

Figure 17. Advective and diffusive transport pathways.

- Assessment of the occurrence of pressure compartments, so that dewatering can be designed to depressurize all compartments significant for slope stability.
- Identification of geologic structures which can aid in slope depressurization.
- Identification of flow barriers.

4.7 Repository performance assessment

Pathways analysis was initially implemented for repository performance assessment, and has been used in repository performance assessment in a number of countries. Repository performance is frequently related to definition of the advective and diffusive transport pathway effective properties and geometry. A conceptual radioactive waste repository design is illustrated in Figure 18. In this design, the primary geosphere barrier is defined by the 100 m scale rock mass between the repository emplacement rooms and fracture zones.

Pathways analysis can provide estimates of key repository performance assessment (PA) measures such as advective pathway length, travel time, pathway dilution, and flux.

Since the fracture pattern is generally known in a purely statistical sense, performance measures may be expressed, for example, in terms of the probability that a flow and transport pathway has travel time T_t less than a critical value T_c. The fracture statistics used for a prototype site are given in Table 1.

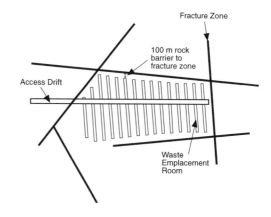

Figure 18. Conceptual repository design

Based on 5000 Monte Carlo simulations, the probability $P[T_t < T_c | R_g, F_g]$ of occurrence of a pathway with a travel time T_t less than $T_c = 10,000$ years, given the repository geometry R_g and the fracture geometry F_g was calculated as 0.36 %. This provides a measure of the reliability of the waste disposal conceptual design based on fracture flow considerations only. This value was calculated through a combination of Monte Carlo simulation and pathways analysis. The analysis can be extended to include, for example, the probability of human intrusion, climate change of seismic/tectonic disruption, and other scenarios.

Table 1: Fracture network statistics.

Parameter		
Spatial Structure	50% terminations at intersections	"BART" Conceptual Model (Dershowitz et al., 1995)
Orientation Distribution	Set 1: Horizontal, Fisher Dispersion K = 25 Set 2: Vertical, Strike N-S, Fisher Dispersion K = 15	Fisher Distribution derived from Terzaghi corrected borehole data
Size	Radius Exponentially Distributed, Mean = 5.3 m	Derived from analysis of tracelengths
Intensity P_{32}	Set 1: 0.43 m^{-1} Set 2: 0.64m^{-1}	as P_{32} fracture surface area per volume (m^2/m^3)
Transmissivity	Lognormally distributed with Mean 10^{-8} m^2/s and standard deviation 10^{-7} m^2/s	Derived from packer tests using OxFilet analysis
Transport Aperture	Related to transmissivity as $e_t = 0.5$ T$^{1/2}$	Doe (1993)

5 CONCLUSIONS

Extensive field evidence demonstrates that discrete, structurally controlled pathways are very important to quantification and understanding of flow and transport in fractured rock masses. For these cases, pathways analysis has a number of advantages over conventional flow and transport modeling (e.g., Dershowitz, 1992):

- Pathway geometry and pathways can be directly quantified based on geological information.
- Computations and engineering effort is significantly less than with conventional DFN or continuum flow and transport modeling.
- Spreading due to network divergence branching and dilution due to network convergence can be considered, where it cannot be modeled by conventional streamline approaches.
- Probabilistic approaches can be incorporated to reflect the uncertainty and variability of conductive and flow barrier geological structures.

The prevalent lognormal or power law distribution of the hydraulic properties of discrete features ensures that there will always exist a few key pathways which dominate flow and transport behavior. Identification of these pathways will lead to better environmental engineering solutions.

ACKNOWLEDGMENTS

We would like to thank the individuals who provided the technical inspiration for this paper, and the organizations which graciously provided permission to cite their work. The individuals include: Scott Donald, Thomas Doe, and Peter Wallmann of Golder Associates, Masahiro Uchida of PNC (Japan), Eugene Wadleigh of Marathon Oil, and Dick Ketelle of Oak Ridge National Laboratory. Much of the work for the technical development of the pathways analysis method was supported by PNC (Japan), and their support is acknowledged with gratitude. Thanks are also due to Marathon Oil and Oak Ridge National Laboratory, and to the Weizmann Institute of Technology (Israel).

REFERENCES

Bear, J, 1972. *Dynamics of Fluids in Porous Media*. Dover Publications, BY.

Dershowitz, W. and R. Busse, 1996. *Discrete Fracture Analysis in Support of the Äspö Tracer Retention Understanding Experiment (TRUE-1)*. SKB Äspö International Cooperation Report. SKB, Stockholm. (in press)

Dershowitz, W. And W. Roberds, 1990. *Strategies for Hydraulic Characterization of a Repository Site in the Opalinus Clay and in the Lower Freshwater Molasse*. Report NIB-90-36. Nagra, Wettingen, Switzerland.

Dershowitz, W., 1992. *The Role of the Stripa Project in the Development of Practical Discrete Fracture Modeling Technology*. Proceedings of the Fourth International NEA/SKB Symposium, Stockholm, 14-16 October. OECD/NEA, Paris.

Dershowitz, W., 1984. *Rock Joint Systems*. Ph.D. Dissertation, MIT, Cambridge, MA.

Dershowitz, W., G. Lee, J. Geler, G. Lee, and P. LaPointe, 1995. *FracMan Interactive Discrete Feature Data Analysis, Geometric Modeling, and Exploration Simulation*. User Documentation. Ver. 2.50. Golder Associates Inc., Seattle.

Einstein, H., 1993. *Modern Developments in Discontinuity Analysis, the Persistence-Connectivity Problem*. Volume 3, Chapter 9 of Hudson et al., Comprehensive Rock Engineering. Pergamon Press, NY.

Doe, T., 1993. *Effective Apertures for Hydraulic Transient Response and Solute Transport*. Unpublished MS. Golder Associates Inc., Seattle.

Doe, T, A. Thomas, P. LaPointe, and J. Thomas, 1996. *Fracture Data Analysis and Flow Modeling of the Kamaishi KD-90 and Task 3-2 Experiments*. PNC Technical Report, PNC, Tokai, Japan.

Forsmark, T and I. Rhen, 1993. *Information for Numerical Modeling*, 1994. General Information and Calibration Cases for the Äspö HRL, Tunnel Section 700-2545 meters. SKB Äspö Project Report PR-25-94-16. SKB, Stockholm.

Horne, R.N., and Rodriguez, F., 1983. *Dispersion in Tracer Flow in Fractured Geothermal Systems*, Geoph. Res. Lett., 10, 289-292.

Kettelle, R.H., R.R. Lee, 1992. *Migration of a Groundwater Contaminant Plane by Stratabound Flow in Waste Area Groupings at Oak Ridge National Laboratory, Oak Ridge Tennessee*, Oak Ridge Laboratory Report ORNLIER-126.

Mazor, E., 1990. Applied Chemical and Isotopic Ground Water Hydrology. John Wiley & Sons, NY.

Ministry of Environment and Energy, 1996. *Landfill Regulations, Ontario Provincial Ministry of Environment and Energy*, Ottawa.

Miller, I., G. Lee, W. Dershowitz, and G. Sharp, 1995. *MAFIC Matrix/Fracture Interaction Code with Solute Transport*. User Documentation, Version 1.5. Golder Associates Inc., Seattle.

Pearson, R, and W. Lozier, 1986. *Tracer Testing at the Bear Creek Site*. Report to Oak Ridge National

Laboratory. Golder Associates Inc., Atlanta.

Roberts, D.L., 1992. NAPSAC *Summary Document*. Atomic Energy Agency (UK), Report AE-D&R-0271.

Robinson, P.C., 1984. *Connectivity, Flow and Transport in Network Models of Fractured Media*. Ph.D. Thesis, Oxford University.

Sedgewick, R., 1988. *Algorithms*. Addison Wesley Publishing, Reading MA.

Sudicky, E. and G. McLaren, 1989. *The Laplace Transform Galerkin Technique: A Time Continuous Finite Element Theory and Application to Mass Transport in Groundwater. Water Resources Research.* Vol. *25,* No 4, pp. 1833-1846.

Uchida, M, W. Dershowitz, A. Sawada, P. Wallmann, and A. Thomas, 1996b. *FracMan Discrete Fracture Modeling for the Äspö Tunnel Drawdown Experiment*. SKB International Cooperation Report ICR-XX. SKB, Stockholm.

Wadleigh, E., *Personal Communication*.

Winberg, A., 1996. *TRUE-1 Radially Converging Tracer Test, Preliminary Results*. SKB, Stockholm.

Miscellaneous

Program
ISRM Board and Council Meetings – August 31 and September 1, 1996
Welcome Party in the underground of Turin – The Railway Link
Banquet at the Galleria di Diana in the Savoia Castle – Venaria Reale
Technical visits
List of exhibitors
List of participants
Author index

Eurock '96, Barla (ed.) © 2000 Balkema, Rotterdam, ISBN 90 5809 843 6

Program

Sunday, September 1, 1996

19.30-20.30   **Welcome reception**

Monday, September 2, 1996

8.00-10.00   Registration
10.00-10.30   **OPENING CEREMONY**
10.30-11.15   "ROCHA MEDAL" Lecture
            Dr. M. P. Board (USA.): "Numerical examination of mining induced seismicity"
11.15-11.45   Coffee Break
11.45-12.30   "SCHLUMBERGER LECTURE AWARD 1996"
            Mr. C. R. Windsor (Australia): "Rock reinforcement systems"

BREAK FOR LUNCH

14.30-15.00   Invited Lecture:
            Prof. R. Goodman (USA): "The foundation of Scott Dam - a case study of
            a complexly heterogeneous rock mass"
15.00-18.30   **WORKSHOP 1**
            Theme 1 - Rock mass modelling: continuum versus discontinuum
            Discussion Leader: C. Fairhurst (USA)
            SCHEDULE
            15.00-15.10   **Introduction**
                            C. Fairhurst
            15.10-15.25   A critical analysis of microstructural modelling
                            C. Boutin
            15.25-15.40   Microstructural modelling of the compressive failure of rock
                            C. Fairhurst
            15.40-15.55   Alternative schemes for the assessment of the equivalent
                            continuum hydraulic properties of rock masses
                            C. Fidelibus, G. Barla, M. Cravero
            15.55-16.10   Continuum modelling of the permeability of fractured rock
                            B. Dershowitz
            16.10-16.30   **Group Discussion**

16.30-17.00	**Coffee Break**
17.00-17.15	A Cosserat continuum model for blocky rock under dynamic loading
	J. Sulem and H. B. Mühlhaus
17.15-17.30	Comparison between continuum and discontinuum modelling of toppling at Chiquicamata Open-Pit Mine, Chile
	M. P. Board
17.30-17.45	Continuum modelling by FLAC of rock bolting on blocky rock
	M. Varona
17.45-18.30	**Group Discussion and Conclusions**

15.00-18.30 **WORKSHOP 2**

Theme 2 - Rock mechanics modelling of large scale natural phenomena

Discussion Leader: L. Richards (New Zealand)

SCHEDULE

15.00-15.15	**Introduction**
	L. Richards
15.15-15.30	**Group Discussion**

Geotechnical Implications of Large Scale Natural Phenomena (LSNP)

- processes and mechanism compared with Small Scale Natural Phenomena (SSNP)
- governing parameters compared with SSNP; current state of knowledge on LSNP
- confidence in prediction and performance of LSNP compared with those for SSNP

Rapid Mass Movements

| 15.30-15.50 | Major movements in New Zealand |

- Mt. Cook rock avalanche
- Green Lake landslide

Current research in NZ on rapid movements

L. Richards

15.50-16.10	Numerical and physical modelling of debris flows
	G. P. Giani, C. Deangeli
16.10-16.30	**Group Discussion**

Thoughts on rock mechanics aspects of LSNP and our current understanding of high speed dynamic processes

| 16.30-17.00 | **Coffee Break** |

Creep Deformation of Mountain Slopes

17.20-18.00	Observations on deep seated gravitational slope deformations
	G. Barla, E. Chiriotti, C. Scavia
18.00-18.30	**Concluding Session**

- How well do we really understand LSNP?
- Do we know enough about these LSNP to justify using modelling techniques for real studies or are these still very much in the realm of pure research?
- What are the priorities for future research?

15.00-18.30	**WORKSHOP 3**
	Theme 3 - Fluid rock interaction in saturated and partially
	saturated media
	Discussion Leaders: F. Santarelli (Italy) and J-C. Roegiers (USA)
	SCHEDULE

15.00-15.10	Introduction
	J-C. Roegiers and F. Santarelli
15.10-15.25	The effect of water on the Ekofisk Chalk
	D. Rhett
15.25-15.40	Subsidence and capillary effects in chalks
	C. Schroeder
15.40-15.55	Rheological modelling of collapsible partially saturated rocks
	E. Papamichos
15.55-16.10	Subsidence induced by water injection in water sensitive
	reservoir rocks: the example of Ekofisk
	V. Maury
16.10-16.30	**Group Discussion**
16.30-17.00	**Coffee Break**
17.00-17.15	Overview of the GRI project on shale behaviour
	F. Mody
17.15-17.30	Conventional and controversial mud selection criteria
	H. Dos Santos
17.30-17.45	The use of cutting to study mud-shale interaction
	R. Holt
17.45-18.00	Modelling of chemico-mechanical coupling around wellbores
	in shales
	M. Dusseault
18.00-18.30	**Group Discussion and Conclusions**

15.00-18.30	**WORKSHOP 4**
	Theme 4 - Risk assessment in rock engineering
	Discussion Leader: H. Einstein (USA)
	SCHEDULE

15.00-15.15	**Introduction - Ground Rules**
	H. Einstein
15.15-15.35	**Formal Debate on Motion 1:**
	Probabilistic approaches and risk assessment must be used
	in rock tunnelling
Proponent:	F. Vuilleumier
Opponent:	W. Steiner
2nd for:	P. Grasso
2nd against:	L. Olsson
15.35-16.00	**Open Discussion on Motion 1**
16.00-16.20	**Formal Debate on Motion 2**
	Probabilistic approaches and risk assessment must be used
	for rock slopes
Proponent:	J. Hudson
Opponent:	M. Panet

2nd for:	L. Bolzanigo
2nd against:	A. Mahtab
16.20-16.45	**Open Discussion on Motion 2**
16.45-17.15	**Coffee Break**
17.15-17.35	**Formal Debate on Motion 3**
	Propabilistic approaches and risk assessment must be used for fluid flow through fractured rock
Proponent:	E. Hakami
Opponent:	O. Stephansson
2nd for:	D. Billaux
2nd against:	J. Muralha
17.35-18.00	**Open Discussion on Motion 3**
18.00-18.30	General Discussion.

END OF DAY 1

19.30 Welcome Reception at the Zapata Station of the Turin Rail Link

Tuesday, September 3, 1996

⇒ **SESSION 1**
Fundamental Aspects in Rock Mechanics and Rock Engineering
Chairman: Prof. C. Fairhurst (USA)

9.00- 9.45 ⇒ Key Note Lecture 1
Prof. R. Nova (Italy): "Mathematical modelling of the mechanical behaviour of soft rocks"

9.45-10.30 ⇒ General Report:
Prof. P. Kaiser (Canada)

10.30-11.00 **Coffee Break**
⇒ **SESSION 2**
Near Surface Rock Engineering
Chairman: Dr. C. Erichsen (Germany)

11.00-11.45 ⇒ Key Note Lecture 2:
Dr. J. Sharp (U.K.): "Present understanding and predictive capability of the stability conditions for very high rock cuts"

11.45-12.30 ⇒ General Report
Prof. P. Egger (Switzerland)

BREAK FOR LUNCH

SESSION 1A - Theme 1 - ORAL PRESENTATIONS
Fundamental Aspects in Rock Mechanics and Rock Engineering
Chairman: J.A. Hudson (U.K.)
Co-Chairman: B.C. Haimson (U.S.A)

Rock properties and testing methods
Extension tests on high porosity chalk
R. Risnes, H. Kaarigstad, Stavanger College (Norway)

Modelling of intact rock
A proposed criterion for rock failure
G. Senyur, Hacettepe University, Ankara (Turkey)

Elasto-plastic behaviour of a weak sandstone
I. Vardoulakis, National Technical University of Athens (Greece)
J. Sulem, A. Oulahna, Ecole Nationale des Ponts et Chaussées/LCPC, Paris (France)
E. Papamichos, T.E. Unander, J. Tronvoll, IKU Petroleum Research, Trondheim (Norway)

Shale behaviour under different test conditions
Effects of brines on mechanical properties of shales under different test conditions
E.F. Sønstebø, P. Horsrud, IKU Petroleum Research, Trondheim (Norway)

Can shale swelling be (easily) controlled?
H.M.R. dos Santos, Petrobras, Rio de Janeiro (Brazil)
A. Diek, J-C. Roegiers, University of Oklahoma, Norman (U.S.A.)
S.A.B. da Fontoura, Catholic University of Rio de Janeiro (Brazil)

15.45   General Discussion
16.15   **Coffee Break**

Shale behaviour under different test conditions
One-dimensional thermal conductivity measurements in quartz-illite and smectitic shales
D. MacGillivray, B. Davidson, M.B. Dusseault, University of Waterloo (Canada)

In situ state of stress
Can inclined exploratory boreholes be used for in situ stress measurements? A review of three case histories
B.C. Haimson, University of Wisconsin (U.S.A.)

Prediction of stress orientation and distribution in Denmark based on borehole breakout data from the Tornquist Fan and Danish Central Graben
M.V.S. Ask, O. Stephansson, Royal Institute of Technology, Stockholm (Sweden)

Hydro-fracturing for stress measurement in intact granitic rock
K. Shin, Central Research Institute of Electric Power Industry, Tokyo (Japan)
F. Li, Institute of Crustal Dynamics, SSB, Beijing (China)
S. Okubo, University of Tokyo (Japan)

Rock mechanical analyses of in-situ stress/strength ratio at the Posiva Oy investigation sites, Kivetty, Olkiluoto, and Romuvaara, in Finland
P.J. Tolppanen, E.J.W. Johansson, Saanio & Riekkola Consulting Engineers, Helsinki (Finland)
J-P. Salo, Posiva Oy, Helsinki (Finland)

18.00   General Discussion

SESSION 1B - Theme 1   -   ORAL PRESENTATIONS
Fundamental Aspects in Rock Mechanics and Rock Engineering
Chairman: N. Barton (Norway)
Co-Chairman: C. Scavia (Italy)

14.30   Madrid Hall

Strength and deformation properties of rock joints
An experimental and analytical study of the roughness of granite joints
L.N. Lamas, Research Officer of LNEC, Lisboa (Portugal)

Prediction of in situ shear strength of rock joints
T.T. Papaliangas, Technological Educational Institution, Thessaloniki (Greece)
A.C. Lumsden, S.R. Hencher, University of Leeds (U.K

Modelling of rock joints
Rock joint shear mechanical bahavior with 3D surfaces morphology and degradation during shear displacement
G. Archambault, R. Flamand, Université du Québec à Chicoutimi, Québec (Canada)
S. Gentier, BRGM, Direction de la recherche, Orléans (France)
J. Riss, Université de Bordeaux,, Talence (France)
C. Sirieix, ANTEA Direction de la géotechnique, Orléans (France)

Modelling of jointed rock masses
Analysis of different-scale deformation studies of rock masses
V.I. Rechitski, Hydroproject Institute, Moscow (Russia)

The strength of rock masses with finite sized joints
M.H. Bagheripour, G. Mostyn, University of New South Wales, Sydney (Australia)

15.45   General Discussion
16.15   **Coffee Break**

16.45   Madrid Hall

Rock mass characterization
Characterization of rock mass structure for tunneling from outcrop survey and comparison with excavation evidence
M. Cravero, G. Iabichino, CNR - Centro Studi per la Fisica delle Rocce e le Geotecnologie, Turin (Italy)
F. Piana, S. Tallone, CNR - Centro Studi sulla Geodinamica delle Catene Collisionali, Turin (Italy)

Techniques to cope with uncertain parameters in geomechanics on different levels of information
D. Wenner, Ingenieurbüro Müller + Hereth, Karlsruhe (Germany)
J.P. Harrison, Imperial College of Science, Technology and Medicine, London (U.K.)

Continuum versus discontinuum modelling
Continuum modeling of fractured rock masses: is it useful?
P.L. La Pointe, P.C. Wallmann, S. Follin, Golder Associates Inc., Redmond (U.S.A.)

DDM modelling of narrow excavations and/or cracks in an anisotropic rock mass
M.A. Kayupov, Institute of Mining, National Academy of Sciences (Kazakhstan)
M. Kuriyagawa, National Institute for Resources and Environment (Japan)

Fracture propagation in rock
Effective growth rules for macro fracture simulation in brittle rock under compression
J.S. Kuijpers, J.A.L. Napier, CSIR Division of Miningtek (South Africa)

18.00   General Discussion

SESSION 2 - Theme 2   -   ORAL PRESENTATIONS
Near Surface Rock Engineering
Chairman: R. Goodman
Co-Chairman: G.P. Giani

14.30   Londra Hall

Rock Mechanics and Rock Engineering in Dams and Foundations
Modelling of arch dams on jointed rock foundations
J.V. Lemos, Laboratório Nacional de Engenharia Civil (LNEC), Lisboa (Portugal)

Stress-and time-dependent deformations in the foundation rock of a gravity dam
T. Rotonda, R. Ribacchi, University of Rome "La Sapienza" (Italy)

Construction in rock at 3550 meters elevation (Jungfraujoch, Switzerland)
W. Steiner, U. Graber, Balzari & Schudel AG, Bern (Switzerland)
H.R. Keusen, Geotest AG, Bern (Switzerland)

Rock Slope Stability
Site Characterization and Analysis
Stability analysis of a jointed rock excavation through a homogenization method
P. de Buhan, S. Maghous, Laboratoire de Mécanique des Solides, ENPC-CERCSO,
Noisy le Grand (France)
A. Bekaert, Bouygues-TP, St-Quentin-Yvelines (France)

An Assessment of the Static and Dynamic Stability of a Neolithic (6,000 Year Old)
Rockfill Burial Mound
R.J. Pine, F. Desaintpaul, University of Exeter (U.K.)
J.C. Sharp, Geo-Engineering, Jersey (U.K.)

16.15   **Coffee Break**

16.45 _____ Madrid Hall

Mining Applications

Stability investigation and reinforcement for slope at Daye Open Pit Mine, China
F. Zhu, Y. Wang, Northeastern University (China)
O. Stephansson, Royal Institute of Technology (Sweden)

Buckling failure at an open-pit coal mine
Ö. Aydan, Tokai University, Shimizu (Japan)
R. Ulusay, Hacettepe University, Ankara (Turkey)
H. Kumsar, Pamukkale University, Denizli (Turkey)
A. Ersen, Australian Consulate General, Istanbul (Turkey)

Large Slope Movements

Numerical modelling of rock slope deformations
J-M. Vengeon, D. Hantz, A. Giraud, D. Ract, University of Joseph Fourier of Grenoble (France)

Tunnelling in Landslides
R. Poisel, A.H. Zettler, Technical University of Vienna (Austria)
W. Unterberger, Geoconsult, Salzburg (Austria)

Shallow Tunnels

Laboratory model tests and numerical analysis of shallow tunnels
D. Sterpi, Kobe University (Japan) and Polytechnic of Milan (Italy)
A. Cividini, Polytechnic of Milan (Italy)
S. Sakurai, Kobe University (Japan)
S. Nishitake, Mitsubishi Heavy Industries (Japan)

18.00   General Discussion

END OF DAY 2

21.00   Banquet at the Sala Diana of the Savoia Castle in Venaria Reale

POSTER SESSIONS

Theme 1: FUNDAMENTAL ASPECTS IN ROCK MECHANICS AND ROCK ENGINEERING

1.1 Rock properties and testing methods
The way of rupture of chalk submitted to a torsional stress
G. Pecqueur, A. Mikolajczak, J.M. Siwak, Ecole des Mines de Douai, (France)

Validation of an anisotropic damage model under torsional loading
G. Pecqueur, A. Mikolajczak, J.M. Siwak, Ecole des Mines de Douai, (France)
A. Dragon, Laboratoire de Mécanique et de Physique des Matériaux, Poitiers (France)

Prediction of anisotropic responses of strength and deformation of schists
M.H. Nasseri, Tarbiat Modaress University, Tehran (Iran)
K.S. Rao, T. Ramamurthy, Indian Institute of Technology, New Delhi (India)

Shear strength of uniform weak sandstone; rock shear tests vs. triaxial compression tests
K. Tani, K. Kudo, Central Research Institute of Electric Power Industry (Japan)
Creep and ultrasonic waves
J.F. Couvreur, J.F. Thimus, Catholoic University of Louvain, Louvain-la-Neuve (Belgium)

Statistical analysis of the influence of geological-mineralogical parameters on the creep behavior of rock salt
I. Plischke, Bundesanstalt für Geowissenschaften und Rohstoffe, Hannover (Germany)

A comparison on the double cantilever beam and short rod fracture toughness test results of Ankara andesite
C. Karpuz, T. Bozdag, Middle East Technical University, Ankara (Turkey)
A statistical method for practical assessment of sawability of rocks
B. Unver, University of Hacettepe, Ankara (Turkey)

1.2 Shale behaviour under different test conditions
Shale strength as function of stress history and diagenesis
M. Gutierrez, G. Vik, T. Berre, Norwegian Geotechnical Institute, Oslo (Norway)

Strength of Pierre I shale as a function of moisture content
J.C. Fooks, M.B. Dusseault, University of Waterloo, Ontario (Canada)

Fluid effects on acoustic wave propagation in shales
R.M. Holt, E.F. Sønstebø, P. Horsrud, IKU Petroleum Research, Trondheim (Norway)
R.M. Holt, R. Skaflestad, NTNU Norwegian University of Science & Technology, Trondheim (Norway)

1.3 Strength and deformation properties of rock joint
Shear behaviour of joint roughness with two types of asperities
H. Kusumi, K. Nishida, T. Suzuki, Kansay University, Osaka (Japan)

Influence of joint roughness on discontinuity shear strength
G. Sfondrini, S. Sterlacchini, University of Milan (Italy)

Influence of particle size on the shear behaviour of rock joints
K.K. Kabeya, CSIR, Miningtek, Pretoria (South Africa)
T.F.H. Legge, Rand Afrikaans University, Johannesburg (South Africa)

1.4 Modelling of intact rock
On the use of artificial neural networks as generic descriptors of geomaterial mechanical behaviour
A. Amorosi, University of Rome "La Sapienza" (Italy)
D.L. Millar, Imperial College of Science, Technology and Medicine, London (U.K.)
S. Rampello, University of Rome "La Sapienza" (Italy)

Derivations of time step magnitudes for self-adaptive viscoplastic implementation of Hoek-Brown failure criterion
I.C. Duarte Azevedo, Federal University of Viçosa, Minas Gerais (Brazil)
E.A. Vargas, L.E. Vaz, Catholic University, Rio de Janeiro (Brazil)

The effect of plane-strain and isotropic loading in hollow-cylinder strength
E. Papamichos, J. Tronvoll, A. Skjærstein, T.E. Unander, IKU Petroleum Research, Trondheim (Norway)
I. Vardoulakis, National Technical University of Athens (Greece)
J. Sulem, Ecole Nationale des Ponts et Chaussées/LCPC, Paris (France)

Effects of induced anisotropy on the strength and deformation behaviour of two sedimentary rocks
A. Pellegrino, AGIP, San Donato Milanese (Italy)
G. Barla, Polytechnic of Turin (Italy)
J. Sulem, Ecole Nationale des Ponts et Chaussées/LCPC, Paris (France)

Development of a brittle rock-like material having different values of porosity, density and strength
V.S.Vutukuri, H. Moomivand, University of New South Wales, Sydney (Australia)

Modeling of partially - saturated collapsible rocks
E. Papamichos, C. Ringstad, IKU Petroleum Research, Trondheim (Norway)
M. Brignoli, F.J. Santarelli, AGIP, Milan (Italy)

1.5 Modelling of rock joints
Prediction of the strength of jointed rock - Theory and practice
M.H. Bagheripour, G. Mostyn, University of New South Wales, Sydney (Australia)

A three-dimensional extension of Amadei-Saeb's 2D rock joints constitutive model
F. Baldoni, ENEL, Roma (Italy)
A. Millard, CEA/DMT, Saclay (France)

Scaling considerations for the modelling of rock joints in laboratory shear tests
J.P. Seidel, C.M. Haberfield, W. Fleuter, Monash University, Melbourne (Australia)

1.6 Modelling of jointed rock masses
Determination of the shear strength factors by the planes of weakness in jointed rock masses
P.A. Fonaryov, Moscow Automobile & Roads Institute (Russia)
V.V. Kononov, Ministry of Construction (Russia)

Predicted and observed behaviour of weakness zones in a hard, jointed rock mass
E. Johansson, Saanio & Riekkola Consulting Engineers, Helsinki (Finland)
J. Pöllä, Technical Research Centre of Finland, Espoo (Finland)
P. Holopainen, City of Helsinky, Geotechnical Division, Helsinki (Finland)

1.7 Rock mass characterization
Development of a system for the automatic construction of discontinuity trace maps and discontinuity measurements from digital images
T.R. Reid, J.P. Harrison, Imperial College of Science, Technology and Medicine (U.K.)

In-situ block size distribution prediction with special reference to discontinuities with fractal spacing distributions
P. Lu, J-P. Latham, University Queen Mary and Westfield College, London (U.K.)

Probabilistic rock mass characterization and deformation analysis
R. Sturk, J. Johansson, H. Stille, Royal Institute of Technology (Sweden)
L. Olsson, Geostatistik (Sweden)

Quantitative study in rock mass around a tunnel using an RVSP/CWS inverse technique
C.X. Wu, M.B. Dusseault, University of Waterloo, Ontario (Canada)

1.8 Continuum versus discontinuum modelling
A Cosserat continuum model for blocky rock under dynamic loading
J. Sulem, Ecole Nationale des Ponts et Chaussées/LCPC, Paris (France)
H.B. Mühlhaus, CSIRO Division of Exploration and Mining, Perth (Australia)

Deformation and failure of crystalline rock and an elastoplastic homogenization theory
Y. Ichikawa, Nagoya University, Nagoya (Japan)
J. Wang, National University of Singapore (Japan)

Critical analysis of microstructural modellings
C. Boutin, Ecole Nationale des Travaux Publics de l'Etat, Vaulx en Velin (France)

Comparative investigations of stress distribution and displacements around shotcrete lined tunnels using the finite-element and the distinct-element method
Th.v. Schmettow, W. Wittke, Technical University of Aachen (Germany)

1.9 In situ state of stress
Experimental verification of the Kaiser effect in rock under different environment conditions
M. Seto, Research Development Corporation of Japan (JRDC), Kawaguchi (Japan)
D.K. Nag, Monash University, Churchill, Victoria (Australia)
V.S. Vutukuri, University of New South Wales, Sydney (Australia)

Rock stress measurements for the underground nuclear waste repository in Finland
A. Öhberg, P. Tolppanen, Saanio & Riekkola Consulting Engineers, Helsinki (Finland)
C. Ljunggren, H. Klasson, Vattenfall Hydropower AB, Lulea (Sweden)

1.10 Fracture propagation in rock
Analysis of the propagation of natural discontinuities in rock bridges
C. Scavia, M. Castelli, Polytechnic of Turin, (Italy)

Fracture surface fractality as a major factor in the evaluation of rock fracture energy
C. Scavia, F. Re, Polytechnic of Turin (Italy)
A. Zaninetti, ENEL-CRIS, Milan (Italy)

A study of the breakdown process in hydraulic fracturing tests conducted in impermeable rocks
D. Garagash, E. Detournay, University of Minnesota (U.S.A.)

1.11 Rock mechanics modelling of large scale natural phenomena

On some exhibitions of structure heterogeneities in rock massif as the effect of large-scale action
N.M. Syrnikov, S.V. Kondratyev, Y.S. Rybnov, Institute for Dynamics of Geospheres, Russian Academy of Sciences, Moscow (Russia)

Experimental analysis of natural shear zones under compressive loading
C. Lamouroux, A. Hedir, D. Kondo, Université de Lille 1, Villeneuve d'Ascq. (France)

Deformation facies of bedding failure in folded rocks
T. Uemura, Niigata

Theme 2: NEAR SURFACE ROCK ENGINEERING

2.1 Rock Mechanics and Rock Engineering in Dams and Foundations

Geophysical and rock mechanics investigations at Berke dam site
T. Bozdag, D. Sari, A.G. Pasamehmetoglu, Middle East Technical University, Ankara (Turkey)

Excavation and support of a rock slope in Sykia damsite, Greece
J. Thanopoulos, D. Dalias, Public Power Corporation, Moyzaki (Greece)

Numerical modelling of a bridge foundation on a jointed rock slope with the distinct element method
Rachez, J-L. Durville, Laboratoire Central des Ponts et Chaussées, Paris (France)

2.2 Rock Slope Stability

Evaluation of sliding instability factor of safety using fuzzy analysis of discontinuity orientation
A.M. Ferrero, Polytechnic of Turin, (Italy)
J.P. Harrison, Imperial College of Science Technology and Medicine, London (U.K.)
G. Scioldo, Geo&Soft Int., Turin (Italy)

Rock mechanics studies to improve intact rock block exploitation and slope stability conditions in a quarry basin
C. Deangeli, O. Del Greco, A.M. Ferrero, G. Pancotti, Polytechnic of Turin (Italy)
G.P. Giani, University of Parma (Italy)

Integrated intelligent modelling on slope stability analysis
Xia-ting Feng, Yong-jia Wang, Northeastern University, Shenyang (China)
K. Katsuyama, National Institute for Resources and Environment, Tsukuba (Japan)

Continuous and discontinuous modelling of the Corvara cliff
F. Lanaro, G. Barla, Polytechnic of Turin (Italy)
L. Jing, O. Stephansson, Royal Institute of Technology, Stockholm (Sweden)

Stabilization of a huge rock block overhanging the inhabited area of Zambana Vecchia (Province of Trento, Italy)
A. Mammino, Rock and Tunnelling Engineer, Treviso (Italy)
L. Cadrobbi, C. Valle, Engineering Geologist, Trento (Italy)
F. Tonon, Structural Engineer, Treviso (Italy)

The identification of bedding shears and their implications for road cutting design and construction
T. Gordon, M.J. Scott, I. Statham, Ove Arup & Partners, Cardiff (U.K.)

2.3 Mining Applications
The use of empirical design techniques to ensure the long term stability and extraction of an orebody at Msauli Asbestos mine
T.J. Kotze, Rock Engineering Consultant - BKS Hatch (South Africa)

Geomechanical scheme and deformations of coal open pit mine slope, measures for stabilizing them
N.H. Hanh, Vietnamese Geotechnical Institute

Appraisal of a systems approach to slope engineering at the Ffos Las opencast coal site
C.P. Nathanail, The Nottingham Trent University, Nottingham (U.K.)

2.4 Large Slope Movements
A numerical interpretation of the observed behaviour of an unstable slope
A. Cividini, Polytechnic of Milan (Italy)
G. Garigali, Consulting Engineer, Milan (Italy)
V. Manassero, Rodio S.p.A., Casalmaiocco, Milan (Italy)

2.5 Shallow Tunnels
Design of shallow tunnel linings
N.N. Fotieva, N.S. Bulychev, A.S. Sammal, Tula State University, Tula (Russia)

Three-dimensional analysis of near-surface tunnels in weak rock
P. Carruba, G. Cortellazzo, University of Padova (Italy)

Wednesday. September 4, 1996

All day technical visits

- Visit A: Turin Underground Railway Link

- Visit B: Underground Power Station and Dam Site, Entracque

- Visit C: Morgex Twin Tunnels along the Aosta-Monte Bianco Highway

- Visit D: Pont Ventoux - Susa Hydroelectric Scheme

- Visit E: Borzoli Cavern (Genova-Voltri Railway Link)

- Visit F: San Pellegrino Road Tunnels

END OF DAY 3

21.00   Concert in the San Filippo Church

Thursday, September 5, 1996

	⇒ **Plenary SESSION 3**
	Rock Engineering at Depth
	Chairman: Prof. P. Berest (France)
9.00- 9.45	⇒ Key Note Lecture 3
	Dr. V. Maury (France): "Specific and compared difficulties in works at great depth in mining, civil engineering and oil industry from a rock mechanics perspective"
9.45-10.30	⇒ General Report:
	Prof. H. Duddeck (Germany)
10.30-10.45	⇒ Updating Report:
	Design of tunnels for the Italian High Speed Railway Lines
	Contribution from Italferr-Sis. T.a.v.
	Dr. M. Palliccia and Dr. P. Pietrantoni (Italy)
10.45-11.15	**Coffee Break**
	⇒ **Plenary SESSION 4**
	Environmental Rock Engineering
	Chairman: Prof. C. Tanimoto (Japan)
11.15-12.00	⇒ Key Note Lecture 4:
	Dr. B. Dershowitz (USA): "Rock mechanics approaches for understanding flow and transport pathways
12.00-12.45	⇒ General Report
	Prof. O. Stephansson (Sweden)

SESSION 3A - Theme 3   -   ORAL PRESENTATIONS
Rock Engineering at depth
Chairman: M. Panet (France)
Co-Chairman: G. Gioda (Italy)

<u>14.30</u>   <u>Congress Hall</u>

Rock Abrasivity and Drillability Roadheaders, Impact Hamers, Hard Rock TBM's
Introducing the "destruction work" as a new rock property of toughness refering to drillability in conventional drill-and blast tunnelling
K. Thuro, G. Spaun, Technical University of Munich (Germany)

Verification of the bidirectional drilling and blasting method using TBM pilot tunnels
Chan-Woo Lee, Technology & Engineering Institute, Kolon Engineering and Construction, Seoul (Korea)
Gyu-Jin Bae, Korea Institute of Construction Technology, Seoul (Korea)
Moon-Kyum Kim, Yonsei University, Seoul (Korea)

Rock Mass Classifications in Tunnelling
Analyzing applicability of existing classification for hard carbonate rock in Mediterranean area
I. Jasarevic, M.S. Kovacevic, Faculty of Civil Engineering, Zagreb (Croatia)

Predictions and Computations for Tunnels and Underground Openings
Recent progress in convergence confinement method
D. Nguyen-Minh, C. Guo, LMS, Ecole Polytechnique, Palaiseau (France)

Study of the behaviour of a petroleum reservoir during depletion using an elasto-plastic constitutive model
P. Longuemare, F. Schneider, A. Onaisi, Institut Français du Pétrol, Rueil-Malmaison (France)
I. Shahrour, Laboratoire de Mécanique de Lille, Villeneuve d'Ascq (France)

15.45   General Discussion
16.15   **Coffee Break**

Rock Reinforcement, Support, and Lining Design
Performance of grouted bolts in squeezing rock
M. Blümel, Technical University Graz (Austria)

16.45   Congress Hall

The mechanical effect of grouted rock bolts on tunnel stability
Y. Jiang, T. Esaki, Kyushu University, Fukuoka (Japan)
Yokota, CTI Engineering Co., Ltd, Fukuoka (Japan)

Large Caverns
Class "A" predictions of the locations of major rock discontinuities at a storage cavern site, using seismic tomography
V.S. Hope, C.R.I. Clayton, University of Surrey, Guildford (U.K.)
G. Barla, Polytechnic of Turin (Italy)

Rock engineering aspects of the underground works for the Guangzhou Pumped Storage Project, China
L. Richards, Consultant in Geo-Engineering (New Zealand)
L. Shaoji, Guangdong Pumped Storage Power Station Joint Venture Corporation (China)
A. Carlsson, T. Olsson, Vattenfall Hydropower AB (Sweden)

Monitoring and Back Analysis
Discontinuum stability analysis of large underground cavern based on borehole survey data and field measurements
J. Picha, C. Tanimoto, K. Kishida, Kyoto University (Japan)
R. Hatamochi, Kansai Electric Power Co. Inc. (Japan)
Kunii, Newjec Inc. (Japan)

18.00   General Discussion

SESSION 3B - Theme 3   -   ORAL PRESENTATIONS
Rock Engineering at depth
Chairman: N. Van Der Merwe (South Africa)
Co-Chairman: R. Ribacchi (Italy)

Wellbore Stability

Laboratory investigation of factors affecting borehole breakouts
H.K. Kutter, H. Rehse, Rock Mechanics Group, Institute for Geology, Ruhr-University Bochum (Germany)

Basin scale rock mechanics and wellbore stability
M. Brignoli, S. Zaho, F. Zausa, D. Giacca, F.J. Santarelli, AGIP, S. Donato M.se, Milano (Italy)

Overstressing, Squeezing Behaviour, Tunnelling at Great Depth

Prediction of squeezing pressure on the Uri Project, Kashmir, India
J. Brantmark, H. Stille, Royal Institute of Technology, Stockholm (Sweden)

Behavior of rock mass during excavation of a twin tube roadway tunnel with pilot tunnel in metamorphic rock formation
C.D. Ou, Public Construction Commission, Taiwan (R.O.C.)
W.C. Chang, National Expressway Engineering Bureau, Taiwan (R.O.C.)

Mining Applications

Measurement and prediction of the rockmass of the Ventersdorp Contact Reef Stopes
G. Güle, CSIR Division of Mining Technology, Johannesburg (South Africa)

15.45   General Discussion
16.15   **Coffee Break**

Mining Applications

The assessment of support characteristics of cemented backfills
L. Gündüz, A. Sentürk, University of Süleyman Demirel, Isparta (Turkey)

Salt Rock Mechanics

Instantaneous plasticity and damage of rocksalt applied to underground structures
L. Thorel, M. Ghoreychi, K. Su, Ecole Polytechnique, Palaiseau (France)

The simulation of the behaviour of underground reservoirs in rock salts including clay partings
V.I. Smirnov, E.M. Shafarenko, Scientifical-Technical Center Podzemgazprom, Moscow (Russia)
N.S. Hachaturjan, International Institute of Geomechanic and Gidrostructures, Moscow, Russia

Induced Seismicity in Deep Mining

Analysis of seismicity and rock deformation during open stoping of a deep orebody
D.A. Beck, B.H.G. Brady, University of Queensland (Australia)
D.R. Grant, Mount Isa Mines Limited, Mount Isa (Australia)

Risk Assessment in Underground Excavations

Predicting hazards in rock engineering using critical mechanism pathway analysis

Y. Jiao, Rock Engineering Consultants, Welwyn Garden City (U.K.) and Institute of Water Conservancy and Hydroelectric Power Research, Beijing (China)

J.A. Hudson, Imperial College, London (U.K.) and Rock Engineering Consultants, Welwyn Garden City (U.K.)

18.00   General Discussion

SESSION 4 - Theme 4   -   ORAL PRESENTATIONS

Environmental Rock Engineering

Chairman: G. Mostyn (Australia)

Co-Chairman: M. Van Sint Jan (Chile)

14.30   Londra Hall

Flow in a single fracture

Investigation of two-phase flow in a single rough fracture

K. Kostakis, J.P. Harrison, Imperial College of Science, Technology and Medicine, London (U.K.)

A.T. Young, British Gas plc. (U.K.)

Fluid Flow in Fractured Rock Masses

An integrated strategy for field data collection, three-dimensional joint representation and modelling of fluid flow in fractured rock masses

J. Hadjigeorgiou, R. Therrien, Université Laval, Quebec (Canada)

In-situ experiment on seepage flow in jointed rock masses and the 3-D seepage flow analysis - Application to radial flow -

K. Kikuchi, Y. Mito, Kyoto University, Kyoto (Japan)

M. Nakata, Mitsui Construction Co. Ltd. (Japan)

Coupled deformation-pore fluid diffusion effects on the development of localized deformation in fault gouge

J.W. Rudnicki, R.J. Finno, M.A. Alarcon, Northwestern University (U.S.A.)

G. Viggiani, Université J. Fourier, Grenoble (Francia)

M.A. Mooney, University of Oklahoma (U.S.A.)

15.45   General Discussion

16.15   **Coffee Break**

Subsidence behaviour of rock structures

Effects on surface structures due to collapsed bord and pillar workings

B.J. Madden, CSIR, Miningtek (South Africa)

J.N. van der Merwe, SASOL Mining PTY Ltd. (Soth Africa)

D.C. Oldroyd, GENCOR Rock Engineering (South Africa)

Nuclear Waste Repositories Rock Damage Zone Characterization
Review of excavation disturbance measurements undertaken within the ZEDEX project:
implications for the Nirex rock characterisation facility
N. Davies, D. Mellor, Nirex Ltd, Oxfordshire (U.K.)

Planning, organization, and execution of an EDZ experiment while excavating two test drifts
by TBM boring and blasting, respectively
O. Olsson, Swedish Nuclear Fuel and Waste Management Co (SKB), Oskarshamn (Sweden)
G. Bäckblom, Swedish Nuclear Fuel and Waste Management Co (SKB), Stockholm (Sweden)
K. Ben Slimane, A. Cournut, ANDRA, Fontenay-aux-Roses (France)
N. Davies, D. Mellor, Nirex Ltd (U.K.)

Study of microcracking of Lac du Bonnet granite
F. Homand-Etienne, A. Sebaibi, Laboratoire de Géomécanique, ENSG, Vandoeuvre-lès-Nancy (France)

Thermal-Hydraulic-Mechanical Coupling
Assessment of thermo-hydro-mechanical interactions for a fractured rock using a finite
difference method
O. Didry, Electricité de France, Direction des Etudes et Recherches, Moret-sur-Loing (France)
K. Su, Ecole Polytechnique, Palaiseau (France)

Predicted and observed heat transfer around a refrigerated rock cavern
R. Glamheden, U. Lindblom, Chalmers University of Technology (Sweden)

18.00   General Discussion
18.30   CLOSING CEREMONY   Congress Hall

POSTER SESSIONS

Theme 3: ROCK ENGINEERING AT DEPTH

3.1 Rock Abrasivity and Drillability Roadheaders, Impact Hamers, Hard Rock TBM's
Forecasting the rock abrasivity in the civil and mining technological fields
N. Innaurato, Polytechnic of Turin (Italy)
R. Mancini, CNR Centro Studio per la Fisica delle Rocce e le Geotecnologie, Turin (Italy)

A model to predict the performance of roadheaders and impact hammers in tunnel drivages
N. Bilgin, S. Yazici, S. Eskikaya, Instanbul Technical University (Turkey)

Prediction of penetration and utilization for hard rock TBMs
M. Alber, Stuttgart (Germany)

The development of rock mass parameters for use in the prediction of tunnel boring machine
performance
C. Laughton, P.P. Nelson, University of Texas at Austin (U.S.A.)

Analysis of rock fragmentation with the use of the theory of fuzzy sets
L.L. Mishnaevsky Jr., S. Schmauder, MPA University of Stuttgart (Germany)

3.2 Wellbore Stability
Deformation and failure of thick-walled hollow cylinders of mudrock - A study of wellbore instability in weak rock
J.R. Marsden, J.W. Dennis, Imperial College, London (U.K.)
B. Wu, CSIRO Division of Petroleum Resources, Glen Waverley (Australia)

Borehole stability in laminated rock
L.N. Germanovich, University of Oklahoma (U.S.A.)
A.N. Galybin, A.V. Dyskin, Univerisity of Western Australia (Australia)
A.N. Mokhel, Russian Academy of Sciences
V. Dunayevsky, Westport Technology Center, IITRI, Texas (U.S.A.)

Modelling a deep borehole excavation in marls using an advanced constitutive model
N. El-Hassan, J. Desrues, R. Chambon, Laboratoire Sols Solides Structures,
UJF-INPG-CNRS, Grenoble (France)

Semi-analytical models for predicting stresses around openings in non-linear geomaterials
P.A. Nawrocki, M.B. Dusseault, University of Waterloo (Canada)
R.K. Bratli, Saga Petroleum A.S., Sandvika (Norway)

Collaborative development of a wellbore stability analysis software with determination of horizontal stress bounds from wellbore data
X. Li, J-C. Roegiers, University of Oklahoma, Norman (U.S.A.)
C.P. Tan, Australian Petroleum Cooperative Research Centre, CSIRO Petroleum,
Melbourne (Australia)

3.3 Rock Mass Classifications in Tunnelling
Actual conditions of an eighty years old, totally unsupported tunnel compared with rock support design using current rock classification systems
L.M. Hansen, Vattenfall Hydropower AB (Sweden)

Correlation between rock mass classes, convergence rates and support densities for underground coal mine excavations
C.D. da Gama, Technical University of Lisbon (Portugal)

3.4 Predictions and Computations for Tunnels and Underground Openings
Effect of abutment compliance on the stability of an underground bedded roof formation
A.I. Sofianos, N.T.U.A. (Greece)
A.P. Kapenis, I.C., U.L. (U.K.)

Numerical simulation of tunnel construction-an assessment of two computational models
G. Beer, H.F. Schweiger, Technical University Graz (Austria)

Opening deformability and lining design in deep tunneling: predictions, remarks and design suggestions
G. Berardi, R. Berardi, University of Genova (Italy)

A random set approach to the uncertainties in rock engineering and tunnel lining design
F. Tonon, Structural Engineer, Treviso (Italy)
A. Mammino, Rock and Tunnelling Engineer, Treviso (Italy)
A. Bernardini, University of Padova (Italy)

A comparative study of the effect of in situ stress field on the stability of underground openings
H. Gerçek, M. Genis, Karaelmas University, Zonguldak (Turkey)

3.5 Rock Reinforcement, Support, and Lining Design
Numerical simulation of reinforcement elements and jointed rock mass interaction
A. Rodino, M. Barale, I.GE.A.S., Torino (Italy)

Design of multi-layer tunnel linings in transversely isotropic rock
N.N. Fotieva, G.B. Kireeva, K.E. Zalessky, Tula State University, Tula (Russia)

Designing of rock supports of tunnels on base of generalized data on thickness of destressed zone
Yu.A. Fishman, V.E. Lavrov, Institute Hydroproject, Moscow (Russia)
A.V. Kolichko, Institute Gidrospetsproekt, Moscow (Russia)

3.6 Large Caverns
Stochastic block theory for the observational construction of large underground opening
Y. Mito, T. Saitoh, Kyoto University, Kyoto (Japan)
I. Hirano, Water Resource Development Public Inc., Saitama (Japan)

3D Modelling for underground excavation at point 1, CERN
A. Sloan, D. Moy, Golder Associates Ltd (U.K.)
D. Kidger, University of Manchester (U.K.)

Using neural networks in rock engineering systems for cavern performance auditing
J. Cai, J. Zhao, Nanyang Technological University (Singapore)
J.A. Hudson, Imperial College, London (U.K.)
X. Wu, University of Illinois, Urbana (U.S.A.)

3.7 Monitoring and Back Analysis
Automated back analysis of ground response in rocks and soils via evolutionary computing
D.L. Millar, Imperial College of Science, Technology and Medicine, London (U.K.)

Performance of Kunimi Tunnel through a fracture zone and its back analysis
M. Sezaki, Miyazaki University (Japan)
Ö. Aydan, Tokay University, Shimizu (Japan)
T. Kamiyama, Miyazaki Prefecture (Japan)
K. Horikoshi, Kumugai Gumi Co., Miyazaki Branch (Japan)

In-situ chamber tests for underground compressed-air storage facilities
Y. Nishimoto, T. Tobase, M. Hori, Electric Power Development Co. Ltd., Tokyo (Japan)
T. Sawada, New Energy Foundation, Tokyo (Japan)

The construction, ground reinforcement, and monitoring of a large cavern in poor rock mass in NW Italy
P. Grasso, K. Rossler, S. Maccan, S. Xu, Geodata, Torino (Italy)

Analysis of the behaviour of a large cavern in a strongly fractured rock mass
D. Faiella, A. Garino, ENEL Construction Division, Torino (Italy)
R. Ribacchi, University of Rome (Italy)

3.8 Overstressing, Squeezing Behaviour, Tunnelling at Great Depth
Advanced monitoring data evaluation for tunnels in poor rock
W. Schubert, A. Steindorfer, Technical University Graz (Austria)

Prediction of in situ stresses based on observations and back analyses
D.R. Brox, H. Hagedorn, Amberg Consulting Engineers Ltd., Regensdorf-Watt (Switzerland)

Modelling of rock temperatures in rock masses
D. Fabre, L. Goy, Université Joseph Fourier, Grenoble (France)

3.9 Mining Applications
Rock engineering investigation into aspects of current mining practice, and proposed new mining method for a Namibian base metal mine
G.J. Sweby, CSIR Division of Mining Technology, Johannesburg (South Africa)
J.W. Klokow, Gold Fields of South Africa Ltd., Johannesburg (South Africa)

The calibration of material models in FLAC for deep gold mine problems
R.W.O. Kersten, Independent consultant, Johannesburg (South Africa)
A.R. Leach, Itasca Africa, Johannesburg (South Africa)

Modelling of large underground cavities reinforced by cable-bolts in different geological and geomechanical configurations
V. Merrien-Soukatchoff, J.P. Piguet, Ecole de Mines de Nancy (France)
D. Thibodeau, INCO (Canada)
F. Wojtkowiak, INERIS, Verneuil en Halatte (France)

Stability analysis of a cemented rockfill pillar at the Pyhäsalmi Mine, Finland
H. Kuula, Helsinki University of Technology (Sweden)

Empirical assessment of failure behaviour of stopes and sill pillars in Swedish cut-and-fill mines
J. Sjöberg, Lulea University of Technology (Sweden)
B. Leijon, Conterra AB, Lulea (Sweden)
S.D. McKinnon, Itasca Consulting Group Inc., Minneapolis (U.S.A.)

3.10 Salt Rock Mechanics

Behavior of sealed solution-mined caverns
P. Berest, B. Brouard, Ecole Polytechnique, Palaiseau (France)
G. Durup, Gaz de France (France)

3D geomechanical simulations of leached cavern, are they mandatory?
T. You, H. Henrion, Géostock, Rueil-Malmaison (France)
G. Vouille, S.M. Tijani, J.N. Farfan, Ecole Nationale Supérieure des Mines de Paris,
Fontainebleau (France)

3.11 Induced Seismicity in Deep Mining

An experimental study of regularities of geological medium deformation for prediction of
mining-induced earthquakes in large-scale mining
A.A. Kozyrev, V.A. Maltsev, V.I. Panin, M.V. Akkuratov, V.V. Zakharov, Mining Institute, Kola
Science Centre, RAS, Apatity (Russia)

Multiparameter seismic risk assessment for deep-level mining
R. Stewart, S. Spottiswoode, CSIR Division of Mining Technology, Johannesburg (South Africa)

Rockburst risk assessment on South African gold mines: an expert system approach
S.J. Webber, CSIR Mining Technolgy, Johannesburg (South Africa)

3.12 Risk Assessment in Underground Excavations

Application of risk assessment methods to underground excavations
R.J. Pine, P.N. Arnold, University of Exeter (U.K.)

Requirements for successful rock engineering prediction in mining
J.N. Van Der Merwe, Sasol Coal Division, Johannesburg (South Africa)

Theme 4: ENVIRONMENTAL ROCK ENGINEERING

4.1 Flow in a single fracture

Design and analysis of an experimental system to measure directional permeabilities of a rock
fracture under normal and shear loading
I.W. Yeo, R.W. Zimmerman, M.H. de Freitas, Imperial College of Science, Technology and
Medicine, London (U.K.)

4.2 Fluid Flow in Fractured Rock Masses

Alternative schemes for the assessment of the equivalent continuum hydraulic properties
of rock masses
C. Fidelibus, CNR - Centro di Studi per le Risorse Idriche e la Salvaguardia del Territorio, Bari (Italy)
G. Barla, Polytechnic of Turin (Italy)
M. Cravero, CNR - Centro di Studi per la Fisica delle Rocce e le Geotecnologie, Turin (Italy)

Interpretation of fracture transmissivity and flow geometry from hydraulic tests
L. Wei, K. Been, Golder Associates Ltd, Nottingham (U.K.)

Measurements in ultra-low permeability Media with time-varying properties
Y. Luo, B. Davidson, M.B. Dusseault, University of Waterloo, Ontario (Canada)

4.3 Subsidence behaviour of rock structures
A statistical approach to the basic inverse problem in land subsidence theory
V.I. Dimova, KULeuven, Heverlee (Belgium)
I.V. Dimov, University of Mining & Geology, Sofia (Bulgaria)

Subsidence and capillary effects in chalks
P. Delage, Y.J. Cui, Ecole Nationale des Ponts et Chaussées (CERMES), Paris (France)
C. Schroeder, Université de Liege (Belgium)

Application of a mixed boundary and finite element method to rock salt caverns numerical analysis
J.E.T.Q. de Menezes, D. Nguyen-Minh, Ecole Polytechnique, Palaiseau (France)

Distinct element modelling and mining induced subsidence: influence of major faults
F. Homand-Etienne, I. Mamane, M. Souley, Laboratoire de Géomécanique, ENSG, Vandoeuvre-lès-Nancy (France)
P. Gaviglio, Université de Besançon (France)
M. Al Heib, Ecole des Mines, Nancy (France)

4.4 Nuclear Waste Repositories Rock Damage Zone Characterization
Integrated characterisation of a rock volume at the Äspö HRL utilised for an EDZ experiment
S.J. Emsley, Nirex Ltd, Harwall, Oxfordshire (U.K.) and Golder Associates, Maidenhead (U.K.)
O. Olsson, R. Stanfors, L. Stenberg, SKB, Äspö Laboratoriet, Figeholm (Sweden)
C. Cosma, Vibrometric OY, Helsinki (Finland)
L. Tunbridge, Norwegian Geotechnical Institute, Oslo (Norway)

Examination of the excavation-disturbed zone in the Swedish ZEDEX tunnel using acoustic emission and ultrasonic velocity measurements
S.D. Falls, R.P. Young, Keele University, Staffordshire (U.K.)

Damage zone characterization in the near field in the Swedish ZEDEX tunnel in situ in situ and laboratory measurements
C. Bauer, F. Homand, Laboratoire de Géomécanique, ENSG, Vandoeuvre-lès-Nancy (France)
K. Ben Slimane, ANDRA, Chatenay-Malabry (France)
K.G. Hinzen, BGR, Hannover (Germany)
S.K. Reamer, ISIS, Overath (Germany)

4.5 Thermal-Hydraulic-Mechanical Coupling
Measurement and an application of thermo-physical properties of rocks
D.A. Gunn, S.T. Horseman, British Geological Survey, Nottingham (U.K.)

The mechanical and Hydraulic characteristics of granite and gneiss under temperature variation
H.S. Lee, H.K. Lee, Seoul National University, Seoul (Korea)
Y.J. Park, K.S. Kwon, Korea Institute of Geology, Mining & Materials (KIGAM), Taejon (Korea)

Thermoporoelastic coupling in hydraulic fracturing
Y. Abousleiman, M. Bai, J.C. Roegiers, University of Oklahoma, Norman (U.S.A.)

4.6 Further Development
Information audits for improving rock engineering prediction, design and performance
J.A. Hudson, Imperial College, University of London (U.K.) and Rock Engineering Consultants, Welwyn Garden City (U.K.)
Y. Jiao, Rock Engineering Consultants, Welwyn Garden City (U.K.) and Institute of Water Conservancy and Hydroelectric Power Research, Beijing (China)

Assessment of the geotechnical properties of the weathered rocks at historical monuments in Korea
H.D. Park, Pai-Chai University, Taejon (Korea)

A new system of a grouting control process using a fuzzy logic approach
A.H. Zettler, R. Poisel, University of Technology, Vienna (Austria)
G. Stadler, INSOND Limited, Neumarkt (Austria)

Tunnelling near the San Pellegrino thermal springs (Italy)
G. Barla, Polytechnic of Turin, (Italy)
M. Naldi, Geodes, Turin (Italy)

Seismic monitoring: improvement in hypocenter location using a doublet technique
E. Fortier, C. Maisons, Géostock, Rueil Malmaison (France)
P. Mechler, Université Pierre et Marie Curie, Paris (France)
M. Valette, Elf-Atochem, Vauvert (France)

The Lingotto Conference Centre - *Centro Congressi*
Eurock '96 Symposium Centre

Entrance to Eurock '96 Symposium Centre

Registration area

A break at the Registration Desks

Partial view of the Exhibition area and of the coffee-
break zone

A view of the main conference hall - *Sala 500*
used for the plenary session

TABLE OF SYMPOSIUM ACTIVITIES

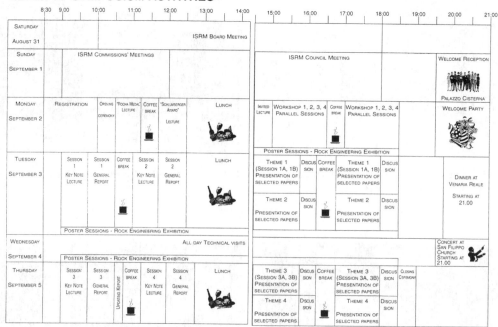

	8:30 – 14:00	15:00 – 21:00
SATURDAY August 31	ISRM Board Meeting	
SUNDAY September 1	ISRM Commissions' Meetings	ISRM Council Meeting — Welcome Reception — Palazzo Cisterna
MONDAY September 2	Registration — Opening Ceremony — "Rocha Medal" Lecture — Coffee Break — "Schlumberger Award" Lecture — Lunch	Invited Lecture — Workshop 1, 2, 3, 4 Parallel Sessions — Coffee Break — Workshop 1, 2, 3, 4 Parallel Sessions — Welcome Party — Poster Sessions - Rock Engineering Exhibition
TUESDAY September 3	Session 1 Key Note Lecture — Session 1 General Report — Coffee Break — Session 2 Key Note Lecture — Session 2 General Report — Lunch — Poster Sessions - Rock Engineering Exhibition	Theme 1 (Session 1A, 1B) Presentation of selected papers — Discussion — Coffee Break — Theme 1 (Session 1A, 1B) Presentation of selected papers — Discussion; Theme 2 Presentation of selected papers — Discussion — Theme 2 Presentation of selected papers — Discussion — Dinner at Venaria Reale Starting at 21.00
WEDNESDAY September 4	All day Technical visits — Poster Sessions - Rock Engineering Exhibition	Concert at San Filippo Church Starting at 21.00
THURSDAY September 5	Session 3 Key Note Lecture — Session 3 General Report — Updating Report — Coffee Break — Session 4 Key Note Lecture — Session 4 General Report — Lunch	Theme 3 (Session 3A, 3B) Presentation of selected papers — Discussion — Coffee Break — Theme 3 (Session 3A, 3B) Presentation of selected papers — Discussion — Closing Ceremony; Theme 4 Presentation of selected papers — Discussion — Theme 4 Presentation of selected papers — Discussion

Torino - La Mole Antonelliana

Eurock '96, Barla (ed.) © 2000 Balkema, Rotterdam, ISBN 90 5809 843 6

ISRM Board and Council Meetings

On August 31 and September 1, 1996 the ISRM Board and ISRM Council meetings took place in the *Unione Industriale* (Industrial Association) building in Turin, where also the ISRM commissions' met on September 1. The overall attendance was very good. The photographs below are intended to report these events.

The ISRM Board Meeting, which took place in the *Salone degli Arazzi* of the *Unione Industriale* (Industrial Association) building in Turin.

The ISRM Council Meeting which took place in the Industrial Association building in Turin.

The ISRM Council Meeting. The head table with President Prof. Sakurai and Secretary General Dr. Rodrigues.

The ISRM Council Meeting. The International Societies table with Prof. Pelizza (ITA President) and Prof. Jamiolkowski (ISSMFE President)

Past President Prof. Franklin paid a visit during the ISRM Board Meeting.

The ISRM Board is relaxing during lunch.

Eurock '96, Barla (ed.) © 2000 Balkema, Rotterdam, ISBN 90 5809 843 6

Welcome party in the underground of Turin – The Railway Link

On Monday September 2, 1996, in the evening, a welcome party took place in the underground of Turin, the *Fermata Zappata* - Zappata station of the Railway Link, presently under construction (see photographs). This complex project, which develops entirely underground, is to reach the two-fold objective of upgrading the city railway system and remodelling the urban environment with the construction of a three-lined central avenue above ground. The welcome address was given by Dr. Emanuela Recchi of the Consortium of Contractors,who is in charge of the project.

Dr. Emanuela Recchi:

On behalf of the RCCF Consortium I am very honoured to welcome you all to our site. We are pleased to have been given the opportunity of showing to the delegates of EUROCK '96 our commitment to the project undertaken by the City Government.

The site project that you will be visiting is an engineering feat not only in terms of magnitude but also for the high technological quality of the methods used in its planning out and its construction.

The Turin railway link is the most important project undertaken by the City since the last war, because it will completely revolutionise the City's railway system. This new order of things will definitely make Turin the true centre of Italy's north-west railway network and will make the city part of the European high-speed network.

The Italian Railway awarded the project to the Consortium of Contractors namely RECCHI (as the main Contractor), CCPL, CIS and FIAT Engineering.

The first phase of the project, the cost of which amounted to 30 million dollars, was completed between 1986 and 1989. From 1990 to 91 the Lingotto, Porta Susa and Zappata station project design was completed. The second phase is now 90% completed and will end this year. The total cost of the second phase is 250 million dollars for civil work and has been financed by both the Italian Railways and the City of Turin.

For this challenging project the Consortium has set up a highly efficient structure for the different activities, which at present involves 300 white and blue collar workers.

Professor Barla's expertise has been indispensable in working out the particulary complex study of the geotechnical aspects.

The City Government, the State Railways and our Consortium had to work closely in order to avoid disrupting normal services and reducing to a minimum the inconveniences to the City's inhabitants.

The consortium's technical know-how and productive capacity have enabled it to fully respect the project's deadlines and costs.

These two aspects are of prime concern as we are fully aware that the Turin Railway Junction is the stepping stone for the new Urban Plan, and a fundamental aspect of the transportation and urbanistic system of Turin in the millenium.

The laying underground of the railway will give rise to an urban conversion project that is the first of its kind in Italy.

All these things considered, we of the Consortium, which today I represent, are proud of the results, and specially glad for this occasion.

In the upper level of Zappata Station we have set up some displays illustrating the project.

Furthermore if you visit the site on Wednesday you will have a more in-depth view of the different techniques and technology used.

Thank you very much for your kind attention,

Turin Railway Link

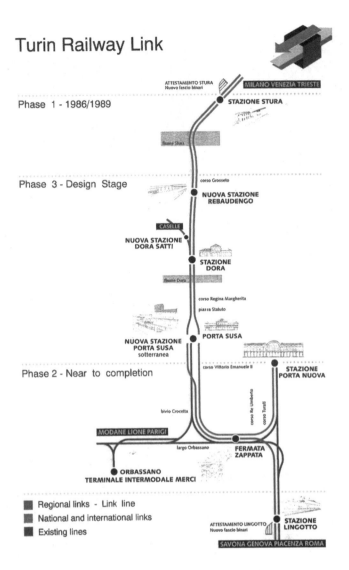

Phase 1 - 1986/1989

Phase 3 - Design Stage

Phase 2 - Near to completion

ATTESTAMENTO STURA
Nuovo fascio binari

MILANO VENEZIA TRIESTE

STAZIONE STURA

fiume Stura

corso Grosseto

NUOVA STAZIONE REBAUDENGO

CASELLE

NUOVA STAZIONE DORA SATTI

STAZIONE DORA

fiume Dora

corso Regina Margherita

piazza Statuto

PORTA SUSA

NUOVA STAZIONE PORTA SUSA sotterranea

corso Vittorio Emanuele II

STAZIONE PORTA NUOVA

bivio Crocetta

corso Re Umberto

corso Turati

MODANE LIONE PARIGI

largo Orbassano

FERMATA ZAPPATA

ORBASSANO TERMINALE INTERMODALE MERCI

STAZIONE LINGOTTO

ATTESTAMENTO LINGOTTO
Nuovo fascio binari

SAVONA GENOVA PIACENZA ROMA

- Regional links - Link line
- National and international links
- Existing lines

A view of the underground railway link. The pilot tunnel in the background.

The participants during the welcome party in the underground of Turin - the Railway Link.

Eurock '96, Barla (ed.) © 2000 Balkema, Rotterdam, ISBN 90 5809 843 6

Banquet at the Galleria di Diana in the Savoia Castle - Venaria Reale

With the intent to give a touch of the beauty of the Royal residencies in *Piemonte*, the banquet took place at the *Galleria di Diana*, in the *Savoia* Castle - *Venaria Reale*, not far from Turin. This beautiful *Galleria* was remodelled in 1716 by Filippo Juvarra on the original structure (1703) due to Michelangelo Garove: it covers a 900 m^2 area, being 15 m in height and 12 m wide. The photographs below are a memory of the place and of the banquet.

The *Galleria di Diana* with the banquet area. This picture was taken prior to the arrival of the Eurock '96 participants.

A panoramic view of the *Galleria di Diana* in the *Savoia* Castle.

A view of the reception which took place
at the entrance of the *Galleria di Diana*,
prior to the banquet.

Flowers are being offered to the ladies, during the banquet.

Eurock '96, Barla (ed.) © 2000 Balkema, Rotterdam, ISBN 90 5809 843 6

Technical visits

Technical visits took place on Wednesday September 4. The program offered to Symposium participants and accompanying persons a total of six visits. A unique feature of the program was the choice to follow-up each technical visit with a tour of selected sites of interest from both the historical and artistic points of view. The following different itineraries (which are briefly described below) were made available (see map).

1. Turin underground Railway Link
2. Underground Power Station and Dam Site, Entracque
3. Morgex Tunnel along the Aosta-Monte Bianco highway
4. Pont Ventoux - Susa Hydroelectric Scheme
5. Borzoli Cavern (Genova-Voltri Railway Link)
6. San Pellegrino Road Tunnels.

The visits were made possible with the help and support of the following:

AEM, Torino
COLLINI, Trento
CONSORZIO L.A.R., Genova
DIPENTA, Roma
ENEL, Torino
GEODES, Torino
CONSORZIO PONT VENTOUX, Roma
RCCF, Torino
RECCHI, Torino

Turin underground Railway Link

The Turin Railway Link, unprecedented for size of work and extent in the city, is to separate medium and long distance railway lines, including high-speed lines. The result is to obtain a transportation system more targeted on demand, so as to be able to meet the requirement of each specific customer.

The "artificial tunnel" excavated by the "cut and cover" method.

The "natural" tunnel.

The railway infrastructure to be built is mostly underground, in either "natural" or "artificial" tunnels, depending on the constraints posed by the operating railway lines in the vicinity, the need to underpass existing railway lines and roads, the proximity of buildings, the interference with pre-existing infrastructures, and services.

The "natural" tunnels (see photographs) were excavated by adopting different systems of ground treatment and pre-support. Jet-grouting techniques in conjunction with cement and non-contaminant silicate injections were used. Excavation took place by using multiple stage excavation by top heading and benching down.

Performance monitoring was carried out during excavation with the purpose to gain in the understanding of the deformational behaviour of the ground surrounding the tunnels and the supporting structures. The purpose has always been to validate the assumed geotechnical parameters of the ground-support interaction schemes adopted for design.

The "artificial" tunnels (see photographs) were constructed by using the "cut and cover" method and diaphragm walls according to a complex system of superposed and/or adjacent tunnels, always excavated without any interruption or interference with the existing double-track line. It is to be noted that the sub-soil conditions in Turin are such that below a certain depth the excavation cannot be carried out by the usual bucket technique and it becomes necessary to use the hydro-cutting technique.

The subsoil conditions in Turin are characterised by the presence of a sand and gravel deposit. Typically this comprises a layer of sand and gravel, ranging from medium to highly dense, down to a depth of 8 to 10 m; below this depth a conglomerate is often present which occurs in lenses.

Underground Power Station and Dam Site, Entracque

The Entracque Power Station complex includes three horizontally oriented prismatic caverns, the machine hall, the valve chamber, and the transformer hall of widths approximately 16 m, 12 m and 16 m respectively. The machine hall has a total height ranging from 39 m to 46 m.

It is a remarkable example of the underground works carried out by ENEL (the Italian Energy Company) between 1970 and 1980. Excavated in a gneiss/schist rock mass, it is definitely one of the first examples of underground works in Italy, where an integrated approach to Rock Engineering was applied, with extensive monitoring during excavation and construction.

The hydroelectric scheme (see the layout below) comprises: (1) the Chiotas Dam (two photographs show the arch-gravity dam during construction, maximum height 130 m), with a $27.3 \cdot 10^6$ m^3 basin, 1980 m asl, (2) the Rovina lake, with a $1.2 \cdot 10^6$ m^3 capacity, 1530 m asl, and (3) the Piastra Dam, with a $9.0 \cdot 10^6$ m^3 capacity, 960 m asl, which is used as a return basin.

Photographs showing the Chiotas arch-gravity dam during construction.

The hydroelectric scheme is in the Gesso Valley, some 80 km from Turin. It is favourably located, being not far from important industrial plants in North Italy. The total installed power is 1500 MW approximately

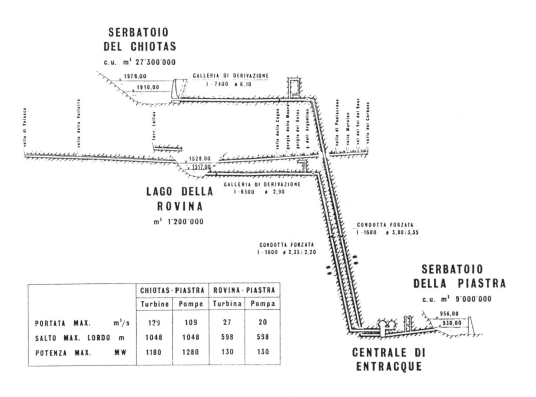

SERBATOIO
DEL CHIOTAS
c.u. m³ 27'300'000

GALLERIA DI DERIVAZIONE
I - 7400 ∅ 6,10

1978.00
1910.00

LAGO DELLA
ROVINA
m³ 1'200'000

GALLERIA DI DERIVAZIONE
I - 6500 ∅ 2,90

1528.00
1517.00

CONDOTTA FORZATA
I - 1000 ∅ 2,35 : 2,20

CONDOTTA FORZATA
I - 1600 ∅ 3,80 : 3,35

SERBATOIO
DELLA PIASTRA
c.u. m³ 9'000'000

956.00
930.00

CENTRALE DI
ENTRACQUE

	CHIOTAS · PIASTRA		ROVINA · PIASTRA	
	Turbine	Pompe	Turbina	Pompa
PORTATA MAX. m³/s	129	109	27	20
SALTO MAX. LORDO m	1048	1048	598	598
POTENZA MAX. MW	1180	1280	130	130

Layout of the hydroelectric scheme.

Morgex Tunnel along the Aosta-Monte Bianco highway

The Morgex tunnel (length approximately 2 km) is being excavated along the Aosta-Monte Bianco highway, presently under construction, with a number of tunnels and viaducts well underway. The main purpose of the highway is to link the existing Torino-Aosta highway to the Monte Bianco tunnel, which takes directly to Chamonix (France) after a length of approximately 11.6 km through the Alps (it is worth mentioning that the Monte Bianco tunnel was excavated between 1959 and 1965).

The Morgex tunnel comprises two parallel tunnels, each one with a cross section of approximately 100-110 m². At present the excavation proceeds from the south-end portals through very poor ground conditions (a "melange" of rock blocks in a clay, sand and gravel matrix) which are expected to last for at least 700 m, prior to reaching a fair to good calcschist rock mass, where excavation will take place by drill and blast. The interest of the visit was centered upon the present use of full-face excavation in very poor ground conditions, which requires the adoption of systematic ground pre-reinforcement and pre-treatment.

As shown in the sketch and photograph below a series of overlapping conical patterns of fibre-reinforced spiles are driven ahead of the tunnel, to stiffen the ground around the perimeter and for an appreciable length ahead of the face. Given the working space which is

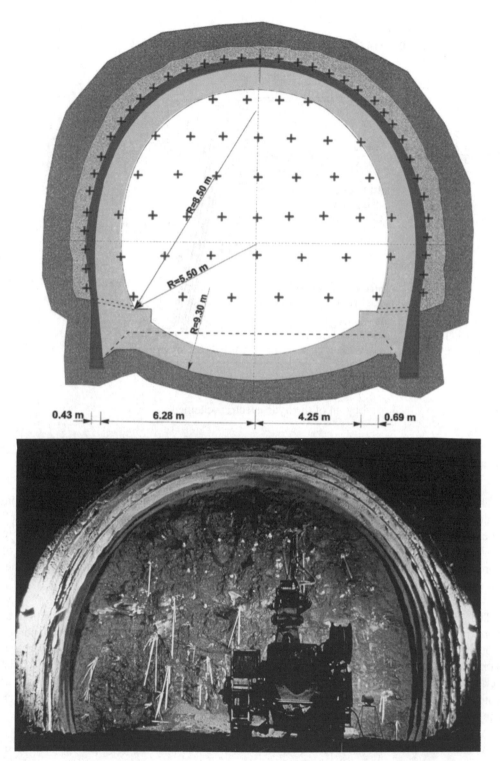

Sketch and photograph showing the full face excavation method with face stabilisation and pre-treatment which is being used successfully at the Morgex tunnel.

made available at the advancing face, large machines can be used for installing support/stabilisation measures at the tunnel perimeter and ahead the face. Monitoring is carried out systematically during excavation and involves monitoring of the ground deformations around the tunnel and in front of the excavation face.

Pont Ventoux-Susa Hydroelectric Complex

The Pont Ventoux-Susa hydroelectric complex is located in the Susa Valley, some 60 km North of Turin. The scheme comprises: the Venaus underground Power Station, with a 150 MW installed power; the intake at the Pont Ventoux location; an offtake free-flow tunnel 14 m long; a daily compensation basin (with a capacity of $560 \bullet 10^6$ m^3) in the Clarea valley; a pressure tunnel 4 km long; a penstock 1.3 km long; two piezometric wells; a return tunnel 1.6 km long; and a return/demodulation basin at the Susa-Gorge, with a capacity of $420 \bullet 10^6$ m^3.

The visit concentrated on the rock mechanics investigations carried out for the design of the underground power house and the access tunnels (see the photographs shown below) which were being excavated in complex geological and hydrogeological conditions, with more than 250 l/s of water being pumped from the main access tunnel. Also visited was the 4.75 m diameter pressure tunnel (see the photographs shown below), being excavated with a TBM. The cavern is located in a calcschist rock mass, mostly of a good quality; the pressure tunnel is in calcschist and gneiss, with a 500 m length in a transition zone, ranging in quality from fair to poor.

Pont Ventoux-Susa hydroelectric complex - Access tunnel.

Pont Ventoux-Susa hydroelectric complex - Pressure tunnel

Borzoli Cavern (Genova-Voltri railway connection)

The Borzoli cavern is part of the Genova-Voltri railway connection between the Valico line (Genova-Milano) and the new Genova-Voltri harbor. In addition to the cavern, a number of tunnels (7 km) are being excavated or are at the design stage (5 km), with a total length of 12 km. The geological conditions are quite complex, with a number of lithologies, grades of metamorphim, and deformation history: schists, dolomite, basalt, serpentinite and shale.

The visit concentrated on the Borzoli cavern. This large cavern (260 m^2 maximum) is being excavated to form the connection between the double-track, main line and the single track, Genova-Ovada line. The interest stems from the very poor rock mass conditions (serpentinite), so as to require extensive use of pre-support/reinforcement methods. The cavern is being excavated in multiple stages (see photograph) involving a typical twin side wall drift method (side wall tunnels, central heading, and benching down). The following provisions are taken during excavation and construction: (1) the outer walls and faces of the two side wall drifts are reinforced using grouted tie-backs for the walls and fiber-glass bars for the face; (2) the crown of the cavern is pre-supported by using the umbrella-arch technique.

Photograph of the Borzoli cavern during excavation.

San Pellegrino Road Tunnels

Two road tunnels (Frasnadello and Antea Tunnels) 1800 m and 400 m long, with a final section of 100 square meters are being excavated for the improving of the regional road along the Brembana Valley. Due to the presence of the well known San Pellegrino mineral water springs in the vicinity of the tunnel alignment, a large-diameter (11.8 m), shielded TBM is being used for excavation (see the photographs below) - the first machine of such a size in Italy. It is to be noted that the excavation of the large tunnel, an exploratory tunnel of 3.9 m diameter was excavated by an open TBM in advance with respect to the enlargement (see the photographs below).

The Frasnadello Tunnel is to meet a sequence of dolomites and limestones, marls and argillites. The Antea tunnel is to be excavated entirely in dolomites. Significant faulting and thrusting affect the area, with some of these faults running westward to border the tectonic wedge of the San Pellegrino springs. The visit gave the opportunity to get acquainted with a very interesting project where a multidisciplinary approach was applied in order to assess the

The primary lining consisting of contiguous pre-cast concrete rings.

geological, hydrogeological and rock mass conditions along the tunnel alignment, and careful monitoring of the hydrogeological parameters was used during excavation.

At the time of the Eurock '96 visit, the large TBM excavation was encountering a number of difficulties through a thrust zone between the dolomite-limestone sequence and the argillites. Here, a lack of sufficient face support, in a saturated ground compound of rock block in a clay matrix, let to a sudden collapse, with the formation of a void above the TBM head. This caused the machine to become blocked. Investigations around the tunnel and ahead of the TBM were being planned in order to take the appropriate actions needed to resume excavation. This is in all cases made possible and easier due to the presence of the pilot tunnel, which was excavated prior to the large tunnel.

The large-diameter TBM is shown at the Frasnadello South Portal.

Eurock '96, Barla (ed.) © 2000 Balkema, Rotterdam, ISBN 90 5809 843 6

List of exhibitors

STAND

ALPETUNNEL GEIE
1091, avenue de la Boisse
73000 Chambéry
France

AUTOSTRADA TORINO-SAVONA SpA
Corso Trieste, 170
10024 Moncalieri (TO)
Italy

C.L.I.O.S. Srl
Viale Fantuzzi, 8
32100 Belluno
Italy

ENEL SpA
DAS - Area Amministrativa
Via G,B, Martini, 3
00198 Roma
Italy

GEOBRUGG, FATZER SA
P.O. Box 239
CH-6595 Riazzino
Switzerland

HARPACEAS
Viale Regina Giovanna, 40
20129 Milano
Italy

ITALFERR - SIS. T.A.V. SpA
Via Marsala, 53/67
00185 Roma
Italy

MTS SYSTEMS GmbH
Hohentwielsteig, 3
14163 Berlin
Germany

POLITECNICO DI TORINO
Corso Duca degli Abruzzi, 24
10129 Torino
Italy

POSTER

ALPETUNNEL GEIE
1091, avenue de la Boisse
73000 Chambéry
France

CONSORZIO IRICAV UNO
Via Tovaglieri, 17
00155 Roma
Italy

GEODATA SpA
Corso Duca degli Abruzzi, 48/e
10129 Torino
Italy

GEODES srl
Corso Galileo Ferraris, 80
10129 Torino
Italy

SEIC SpA
Via Torino, 34
34123 Trieste
Italy

SISGEO srl
Via Morandi, 27
20090 Segrate (MI)
Italy

VERLAG GLUECKAUF GmbH
Montebruchstr., 2
D-45219 Essen
Germany

BROCHURE

ALPETUNNEL GEIE
1091, avenue de la Boisse
73000 Chambéry
France

AMBERG MEASURING TECHNIQUE LTD
Trockenloostrasse, 21 - P.O. Box 27
CH-8105 Regensdorf-Watt
Switzerland

CHAPMAN & HALL
2-6 Boundary Row
SE1 8HN London
England

C.L.I.O.S. srl
Viale Fantuzzi, 8
32100 Belluno
Italy

GEOPHYSIK GGD MBH
Bauznerstr., 67
D-04332 Leipzig
Germany

GEOSTUDI ASTIER snc
Via Riccardo Zandonia, 11
00194 Roma
Italy

HARPACEAS
Viale Regina Giovanna, 40
20129 Milano
Italy

SEIC SpA
Via Torino, 34
34123 Trieste
Italy

Partial view of the technical exhibition and poster session areas.

Eurock '96, Barla (ed.) © 2000 Balkema, Rotterdam, ISBN 90 5809 843 6

List of participants

Beck David
University of Queensland
Dept. of Mining Minerals & Materials
QLD Brisbane
AUSTRALIA

Lane Christopher
Soil & Rock Engineering Pty Ltd
Suite 11 41 Walters Drive
Herdsman - 6016 Perth
AUSTRALIA

Lee Maxwell
Australian Mining Consultants Pty Ltd
3/15 Queens Street
3000 Melbourne
AUSTRALIA

Maconochie A. Paul
CSIRO
P.O. Box 883
4069 Kenmore
AUSTRALIA

Mostyn Garry
University of New South Wales
School of Civil Engineering
2052 Sydney
AUSTRALIA

Nag Dilip
Monash University
School of Engineering, Switchback Road
3842 Churchill
AUSTRALIA

O'Rourke Greg
Fluor Daniel PTY Ltd
P.O. Box 7016
6850 Cloisters Square
AUSTRALIA

Seidel Julian
Monash University
Wellington Road
3168 Clayton
AUSTRALIA

Walton Robert
Mindata Australia Pty Ltd
115 Seaford Road
3193 Seaford
AUSTRALIA

Windsor Christopher
Rock Technology Research
P.O. Box 1605
6904 Subiaco WA
AUSTRALIA

Blumel Manfred
TU Graz
Inst. fuer Felsmechanik und Tunnelbau
Rechbauerstrasse 4, 8010 Graz
AUSTRIA

Goldschmidt Ernst
Bundesministerium Fuer Landesverteidigung
Rossauer Laende 1
1090 Wien
AUSTRIA

Kurka Heinz
Bundesministerium fuer Landesverteidigung
Rossauer Laende 1
1090 Wien
AUSTRIA

Melchart Roland
Bundesministerium Fuer Landesverteidigung
Rassauer Laende 1
1090 Wien
AUSTRIA

Melchart Roland
Bundesministerium Fuer Landesverteidigung
Rassauer Laende 1
1090 Wien
AUSTRIA

Poisel Rainer
Technische Universitaet Wien
Inst. fuer Geologie, Karlsplatz 13
1040 Wien
AUSTRIA

Schubert Wulf
TU Graz, Institut fuer Felsmechanik u. Tunnelbau
Rechbauerstrasse 12
8160 Graz
AUSTRIA

Schweiger Helmut F.
TU Graz, Institute for Soil Mechanics & Foundation
Engineering - Rechbauerstrasse 12
8010 Graz
AUSTRIA

Steindorfer Albert
TU Graz
Institut fuer Felsmechanik U. Tunnelbau
Rechbauerstrasse 12
8010 Graz
AUSTRIA

Zettler Alfred
Technical University of Vienna
Institute of Geology
Karlsplatz 13/203
1040 Wien
AUSTRIA

Couvreur Jean Francois
Université Catholique Louvaine
Unité de Génie Civil
Place du Levant 1
1348 Louvaine la Neuve
BELGIUM

Schroeder Christian
University of Liege-LGIH
Sart Tilman B19
4000 Liege
BELGIUM

Vervoort Andre
Kuleuven
W. de Croylaan 2
3001 Heverlee
BELGIUM

Azevedo Izabel
Federal University of Vicosa
Condominio Acamari 24
36570 Vicosa
BRAZIL

Ayres da Silva Lineu
Escola Politecnica, Univ. of Sao Paulo, Minign
Dept.
Rua Bernardino de Campos 982
04620-003 Sao Paulo
BRAZIL

Fontoura Sergio
Puc Rio
Rua Marques de S. Vicente 225
Rio de Janeiro
BRASIL

Quadros Eda F.
Instituto de PTS
Quisas Tecnologicas de Sao Paulo
Caixa Postal 7141
01064-970 Sao Paulo
BRAZIL

Archambault Guy
Université du Quebec
a Chicoutimi
555, Bld. de l'Université
G7H2BI Chicoutimi Quebec
CANADA

Dusseault Maurice
University of Waterloo
Porous Media Research Inst.
Dept. of Earth Sciences
N2L 3G1 Waterloo ON
CANADA

Fooks Jeanette
University of Waterloo
211511 Albert Street
N2L 5A7 Waterloo ON
CANADA

Franklin John
Franklin Geotechnical, Wipware Inc.
307162 Hockley Rd.
L9W 2Y8 Orangeville ON
CANADA

Hadjigeorgiou John
Université Laval
Dept. of Mining
GIK7P4 Quebec
CANADA

Kaiser Peter K.
GRC Laurentian University
Ramsey Lake Road
P3E2C6 Suobury
CANADA

Mac Gillivray David
University of Waterloo
200 University Avenue W.
Waterloo Ontario
CANADA

Martin Derek
Geomechanics Research Centre
Laurentian University
Ramsey Lake Road
P3E2C6 Sudbury ON
CANADA

Merino Luis
Ingenieria de Rocas Ltda.
Fidel Oteiza 1921
Of. 1003 Providencia
Santiago
CHILE

Tapia René
Codelco Chile
Divisiòn Salvador
Av.da Bernardo O'Higgins 103
El Salvador
CHILE

Van Sint Jan Michel
Universidad Catolica Chile
Casilla 306
Santiago 22
CHILE

Fu Bingjun
Chinese Society for Rock Mechanics & Engineering
P.O. Box 9701
100101 Beijing
CHINA

Konecny Pavel
Institute of Geonics
Institut Geoniky AVCR Studentska 1768
70800 Ostrava
CZECH REPUBLIC

Takla Georges
DPB AS
73521 Paskov
CZECH REPUBLIC

Holecko Josef
DPB A.S.
73921 Paskov
CZECH REPUBLIC

Cosam Calin
Vibrometric
Taipaleentie 127
01860 Perttula
FINLAND

Johansson Erik
Saanio & Riekkola Consulting Engineers
Laulukuja 4
00420 Helsinki
FINLAND

Kuula Harri
Helsinki University Technology
Vuorimiehentie 2
02150 Espoo
FINLAND

Ohberg Antti
Saanio & Riekkola Consulting Engineers
Laulukuja 4
00420 Helsinki
FINLAND

Tolppanen Pasi
Saanio & Riekkola Consulting Engineers
Laulukuja 4
00420 Helsinki
FINLAND

Berest Pierre
Ecole Polytechnique
Route de Saclay
91128 Palaiseau
FRANCE

Cuisiat Fabrice
Inst. of Protection & Nuclear Safety
IPSN DES/SESID BP6
92265 Fontenay aux Roses
FRANCE

De Buhan Patrick
ENPC CERCSO
Central 2 La Courtine
93167 Noisy Le Grand
FRANCE

Desaintpaul Florent
School of Mines
4310 Route de la Vitarelle
82000 Montauban
FRANCE

Durville Jean Louis
LCPC
58 Bld Lefebvre
75015 Paris
FRANCE

El Hassan Nada
Université J. Fourier
Lab. 3S - Sols Solides Structures, BP 53X
38041 Grenoble Cedex
FRANCE

Fabre Denis
Université J. Fourier
IRIGM - B.P. 53
38041 Grenoble Cedex 9
FRANCE

Fortier Eric
Geostock
7 Rue E. et A. Peugeot
92563 Rueil Malmaison
FRANCE

Fourmaintraux Dominique
ELF EAP
CSTJF L1136
64018 Pau
FRANCE

Hantz Didier
Université de Grenoble
Lab. de Géologie et Mécanique
B.P. 53 Cedex 9
38041 Grenoble
FRANCE

Homand Francoise
ENSG INPL
Lab. de Geomecanique
Rue du Doyen M. Roubault
B.P. 40
54501 Vandoeuvre les Nancy
FRANCE

Lamouroux Christian
Université des Sciences et Technologies de Lille
UFR Sciences de la Terre
59655 Villeneuve d'Ascq Cedex
FRANCE

Longuemare Pascal
Institute Francais du Pétrol
1-4 Avenue de Bois-Préau
B.P. 311
92506 Rueil Malmaison
FRANCE

Maghous Samir
ENPC CERCSO
Central 2
La Courtine
93167 Noisy Le Grand
FRANCE

Maisons Christophe
Geostock
7 Rue E. et A. Peugeot
92563 Rueil Malmaison
FRANCE

Maury Vincent
ELF Aquitaine
64018 Pau
FRANCE

Millard Alain
CEA/DMT/LAMS
CEN Saclay
91191 Gif/Yvette
FRANCE

Nguyen Minh Duc
LMS, Ecole Polytechnique
91128 Palaiseau
FRANCE

Onaisi Atef
Total, CST Domaine de Beauplan
Route de Versailles
78470 St. Rémy les Chevreuse
FRANCE

Panet Marc
Groupe Simecsol
8 Avenue Newton
92350 Le Plessis Robinson
FRANCE

Pecqueur Guillaume
Mines de Douai
941 Rue Charles Bourseul
BP 838
59508 Douai
FRANCE

Rachez Xavier
LCPC
58 Bld. Lefebvre
75015 Paris
FRANCE

Shao Jianfu
Lab. Mecanique de Lille
Ura 1441 CNRS
Bd. Paul Langevin
59655 Villeneuve d'Asq
FRANCE

Su Kun
G.3S, Ecole Polytechnique
91128 Palaiseau
FRANCE

Sulem Jean
ENPC
93167 Noisy le Grand Cédex
FRANCE

Sultan Nabil
ENPC-CERMES
La Courtine
93167 Noisy le Grand
FRANCE

Thorel Luc
G.3S, Ecole Polytechnique
Route de Saclay
91128 Palaiseau
FRANCE

Vengeon Jean-Marc
Université de Grenoble - Laboratoire Geologie et
Mécanique - BP 53 Cedex 9
38041 Grenoble
FRANCE

Viggiani Gioacchino
Université J. Fourier, Laboratoire 3S
Domaine Universitaire
BP 53X, 38041 Grenoble
FRANCE

Vouille Gerard
Ecole des Mines de Paris
35 Rue Saint Honoré
77305 Fontainebleau Cedex
FRANCE

Wojtkowiak Francis
INERIS, Dir. Scientifique et de la Qualité
Parc Technologique Alata - B.P. 2
60550 Verneuil en Halatte
FRANCE

You Thierry
Geostock
7 Rue E. et A. Peugeot
92563 Rueil Malmaison
FRANCE

Alber Michael
Essigweg 7
70565 Stuttgart
GERMANY

Alheid Hans Joachim
Federal Inst. of Geosciences
Stilleweg 2
30655 Hannover
GERMANY

Bork Winfried
MTS Systems GmbH
Hohentwielsteig 3
14163 Berlin
GERMANY

Duddeck Heinz
TU Braunschweig
Inst. fuer Statik
Beethovenstrasse 51
38106 Braunschweig
GERMANY

Erichsen Claus
WBI
Henricistrasse 50
52072 Aachen
GERMANY

Heinrich Friedhelm
TU BergakademieFreiberg
Zeuenerstrasse 1
09596 Freiberg
GERMANY

Kutter Herbert K.
Institut fuer Geologie
Ruhr Universitaet Bochum
Universitaetstrasse 150
44780 Bochum
GERMANY

Kutter Herbert K.
Institut fuer Geologie
Ruhr Universitaet Bochum
Universitaetstrasse 150
44780 Bochum
GERMANY

Plischke Ingo
Bundesanstalt Geowissenschaften
Stilleweg 2
30655 Hannover
GERMANY

Quick Hubert
Ingenieursozietaet
Katzenbach und Quick
Pfaffenwiese 36
65931 Frankfurt Am Main
GERMANY

Rosetz Gitz P.
TU Freiberg, Inst. fuer Geotechnik
Zennerstrasse 1
09596 Freiberg
GERMANY

Spaun Georg
TU Munchen Lehrstuhl Allgemeine
Agewandte und ingenieur geologie
Lichtenbergstrasse 4
85748 Garching
GERMANY

Stark Alfred
Universitaet Hannover
Institut fuer Unterirdisches Bauen
Welfengarten 1
30167 Hannover
GERMANY

Staudtmeister Kurt
Universitaet Hannover
Institut fuer Unterirdisches Bauen
Welfengarten 1
30167 Hannover
GERMANY

Stoetzner Ulrich
Geophysik GGD
Postfach 241216
04332 Leipzig
GERMANY

Van Nguyen Hong
Universitaet Hannover
Schulenburger Landstrasse 27
30165 Hannover
GERMANY

Wenner Dieter
Mueller+Hereth,
Ingenieurburo fuer Tunnel Felsbau
Am Sandfeld 15A
76149 Karlsruhe
GERMANY

Karlaftis Aristides
A.D.K. Consulting Engineers
Themistokleous 106
10681 Athens
GREECE

Papaliangas Theodossios
Technology Educational Institution
P.O. Box 14561
54101 Thessalonici
GREECE

Sofianos Alexandros
N.T.U.A.
Dept. of Mining Engineering
163 Tritis Septembrion Str.
11251 Athens
GREECE

Thanopoulos Yannis
Public Power Corporation
9 Asteriapou Str.
41222 Larissa
GREECE

Nasseri Mohammad Hossein
University of Tarbiat Modares
Dept. of Mining Engineering
Faculty of Engineering
P.O. Box 14155
4838 Tehran
IRAN

Amorosi Angelo
Università di Roma
Dip. Ingegneria Strutturale
Via Monte d'Oro 28
00186 Roma
ITALY

Angelino Claudio
Via Millio 41
10141 Torino
ITALY

Apuani Tiziana
Università di Milano
Via Mangiagalli 34
20133 Milano
ITALY

Barbero Monica
Politecnico di Torino
Dip. Ingegneria Strutturale
C.so Duca degli Abruzzi 24
10129 Torino
ITALY

Baldovin Giuseppe
Studio Tecnico G. Baldovin
Via Roncaglia 14
20146 Milano
ITALY

Baldovin Ezio
Geotecna Progetti S.p.A.
Via Roncaglia 14
20146 Milano
ITALY

Barla Giovanni
Politecnico di Torino
Dip. Ingegneria Strutturale
C.so Duca degli Abruzzi 24
10129 Torino
ITALY

Biscontin Giovanna
Via Baruzzi 13
35129 Padova
ITALY

Brignoli Marco
Agip. S.p.A.
C.P. 12069
20120 Milano
ITALY

Belletti Paolo
RCT S.r.l.
Via G. di Vittorio 2
20060 Liscate MI
ITALY

Beneforti Stefano
Via Cesare Battisti 19
20122 Milano
ITALY

Berardi Riccardo
Università di Genova, Istituto di Scienza delle
Costruzioni, Via Montallegro 1
16145 Genova
ITALY

Biscontin Giovanna
Via Baruzzi 13
35129 Padova
ITALY

Borgna Stefania
Politecnico di Torino, Dip. Ingegneria Strutturale
C.so Duca degli Abruzzi 24
10129 Torino
ITALY

Borri Brunetto Mauro
Politecnico di Torino, Dip. Ingegneria Strutturale
C.so Duca degli Abruzzi 24
10129 Torino
ITALY

Cabella Enrico
Università di Genova, Istituto Scienza delle
Costruzioni, Via Montallegro 1
16145 Genova
ITALY

Caldara Mario
Edison S.p.A.
Via Rosellini 15/17
20124 Milano
ITALY

Cancelli Andrea
Università di Milano
Dip. Scienze della Terra
Via Sansovino 14
20133 Milano
ITALY

Cargnel Giannangelo
Viale Fantuzzi 8
32100 Belluno
ITALY

Carminati Stefano
Eniricerche S.p.A.
Via Maritano 26
20097 San Donato Milanese MI
ITALY

Cassinis Carlo
Via Chiana 5
00198 Roma
ITALY

Castelli Marta
Politecnico di Torino, Dip. Ingegneria Strutturale
C.so Duca degli Abruzzi 24
10129 Torino
ITALY

Carrubba Paolo
Università di Padova
Via Ognissanti 39
35129 Padova
ITALY

Chiari Antonio
Autostrada Torino Savona S.p.A.
Corso Trieste 170
10024 Moncalieri TO
ITALY

Chiriotti Elena
Politecnico di Torino, Dip. Ingegneria Strutturale
C.so Duca degli Abruzzi 24
10129 Torino
ITALY

Cividini Annamaria
Politecnico di Milano
Piazza Leonardo da Vinci 32
20133 Milano
ITALY

Coli Massimo
Dip. Scienze della Terra
Via G. La Pira 4
50121 Firenze
ITALY

Corona Pietro
Studio Corona
Corso Re Umberto 23
10128 Torino
ITALY

Cotti Italo
Università di Roma "La Sapienza"
Via A. Gramsci 53
00197 Roma
ITALY

Cravero Masantonio
CNR Fisica Rocce e Geotecnologie
C.so Duca degli Abruzzi 24
10129 Torino
ITALY

Crosta Giovanni
Università degli Studi di Milano
Via Mangiagalli 34
20133 Milano
ITALY

Cotecchia Vincenzo
Istituto di Geologia Applicata e Geotecnica
Politecnico di Bari, Via E. Orabona 4
70125 Bari
ITALY

Cuzzani Maria Grazia
Via Bainsizza 16
40133 Bologna
ITALY

De Col Raffaele
Provincia Autonoma di Trento
Servizio Geologico
Via Vannetti 41
38100 Trento
ITALY

De Donati Alberto
Unitalc S.p.A.
Via Nazionale 4
23010 Postalesio SO
ITALY

De Gasperis Giuseppe
Luzenac Val Chisone
Corso Torino 364/366
10064 Pinerolo TO
ITALY

Deangeli Chiara
Politecnico di Torino
Dip. Georisorse e Territorio
C.so Duca degli Abruzzi 24
10129 Torino
ITALY

D'Elia Beniamino
Università di Roma "La Sapienza"
Dip. Ingegneria Strutturale e Geotecnica
Via A. Gramsci 53
00197 Roma
ITALY

Di Maio Sergio
A.G.I.
Piazza Bologna 22
00162 Roma
ITALY

Esu Franco
Università di Roma "La Sapienza"
Dip. Idraulica Trasporti e Strade
Via Eudossiana 18
00184 Roma
ITALY

Fagiani Paolo
Aquater S.p.A.
Via Miralbello 53
61047 S. Lorenzo in Campo PS
ITALY

Faiella Domenico
ENEL S.p.A. DCO
Via G.B. Martini 3
00198 Roma
ITALY

Fava Adriano
Alpina
Via Ripamonti 2
20136 Milano
ITALY

Ferrero Annamaria
Politecnico di Torino
Dip. Georisorse e Territorio
C.so Duca degli Abruzzi 24
10129 Torino
ITALY

Ferrero Cesare
L.G.L. S.n.c.
Via Servettaz 17/2
17100 Savona
ITALY

Fidelibus Corrado
CNR CERIST - Politecnico di Bari
Ist Geologia Applicata e Geotecnica
Via E. Orabona 4
70125 Bari
ITALY

Filippi Dario
L.G.L. S.n.c.
Via Servettaz 17/2
17100 Savona
ITALY

Gallarà Franco
Alpetunnel
Via Sacchi 4
10125 Torino
ITALY

Gambelli Claudio
Italferr
Via Lamaro 13
00173 Roma
ITALY

Garino Antonio
ENEL S.p.A. DCO/TO
Via Avogadro 30
10121 Torino
ITALY

Giani Gian Paolo
Università di Parma, Facoltà di Ingegneria
Viale delle Scienze Campus
43100 Parma
ITALY

Giglio Ignazio
Regione Siciliana 2551
90145 Palermo
ITALY

Grasso Piergiorgio
Geodata S.p.A.
C.so Duca degli Abruzzi 48/E
10129 Torino
ITALY

Griffini Lamberto
Via E. Pagliano 37
20149 Milano
ITALY

Iabichino Giorgio
CNR - Centro Studi Fisica Rocce e Geotecnologie
C.so Duca degli Abruzzi 24
10129 Torino
ITALY

Iannotta Fausto
Italferr - Gruppo FS
Via Lamaro 13
00173 Roma
ITALY

Innaurato Nicola
Politecnico di Torino
Dip. Georisorse e Territorio
C.so Duca degli Abruzzi 24
10129 Torino
ITALY

Lembo Fazio Albino
Terza Università di Roma
Dip. Scienze Ingegneria Civile
Via Corrado Segré 60
00146 Roma
ITALY

Luccioli Paolo
Università di Perugia
Facoltà di Ingegneria
Via XX Settembre 89
06034 Foligno PG
ITALY

Mahtab Ashraf
Geodata S.p.A.
C.so Duca degli Abruzzi 48/E
10129 Torino
ITALY

Mammino Armando
S.I.GE.S. S.a.s.
Via Povegliano 10
Loc. Camalò di Povegliano
31050 Treviso
ITALY

Maranini Enrico
Università di Ferrara
Dip. Scienze Geologiche e Paleontologiche
C.so Ercole I d'Este 32
44100 Ferrara
ITALY

Marcellino Paolo
S.G.I.
Via Ripamonti 89
20139 Milano
ITALY

Marchese Francesco
Italferr Gruppo FS
Via Lamaro 13
00173 Roma
ITALY

Marchisella Raffaele
Italferr SIS T.A.V.
Via Lamaro 13
00173 Roma
ITALY

Martinetti Sandro
ENEL DCO
Via G.B. Martini 3
00198 Roma
ITALY

Mascia Giovanni
Via Boiardo 14
09047 Selargius CA
ITALY

Mazzoccola Daria
Università di Milano, Dip. Scienze della Terra
Via Mangiagalli 34
20133 Milano
ITALY

Meardi Pietro
Via Castelmorrone 4
20129 Milano
ITALY

Miliziano Salvatore
Università di Roma
Dip. Ing. Strutturale e Geotecnica
Via Monte d'Oro 28
00186 Roma
ITALY

Minimax s.a.s.
dott. Marcello
Via P. Canal 12/1
35137 Padova
ITALY

Modugno Chiara
Dip. Scienze della Terra
Via G. La Pira 4
50121 Firenze
ITALY

Naldi Mario
Geodes S.r.l.
Corso Galileo Ferraris 80
10129 Torino
ITALY

Nosetto Andrea
Aquater S.p.A.
Via Miralbello 53
61032 S. Lorenzo in Campo PS
ITALY

Nova Roberto
Politecnico di Milano
P.zza Leonardo da Vinci 32
20133 Milano
ITALY

Oreste Pierpaolo
Viale Cappuccini 10
10023 Chieri TO
ITALY

Paviani Irene
Via Vallona 27
48100 Ravenna
ITALY

Peila Daniele
Politecnico di Torino
TUSC- Dip. Georisorse e Territorio
Corso Duca degli Abruzzi 24
10129 Torino
ITALY

Pellegrino Antonio
Agip S.p.A.
C.P. 12069
20120 Milano
ITALY

Pittaluga Federico
Università di Genova, Dip. Scienze della Terra
Corso Europa 26
16132 Genova
ITALY

Pizzarotti Enrico Maria
S.IN.C. S.r.l.
Via dei Cignoli 9
20151 Milano
ITALY

Poggio Massimo
Injectosond Italia S.r.l.
Via T. Invrea 9/4
16129 Genova
ITALY

Re Fabio
Politecnico di Torino,Dip. Ingegneria Strutturale
C.so Duca degli Abruzzi 24
10129 Torino
ITALY

Rea Giovanni
Italferr SIS T.A.V.
Via Lamaro 13
00173 Roma
ITALY

Ribacchi Renato
Università di Roma
Dip. Ingegneria Strutturale e Geotecnica
Via Monte d'Oro 28
00186 Roma
ITALY

Ricca Armando
Alpina S.p.A.
Via Ripamonti 2
20136 Milano
ITALY

Rodino Alessandro
I.GE.A.S. Ingg. Rodino e Barale
Via Reduzzi 9
10134 Torino
ITALY

Romani Giuseppe
Gruppo Dipenta Costruzioni S.p.A.
Via Agrigento 5
00161 Roma
ITALY

Rondena Enrico
ENEL S.p.A. CRIS
Via Ornato 90/14
20162 Milano
ITALY

Rossi Pier Paolo
ISMES S.p.A.
Via Pastrengo 9
24068 Seriate BG
ITALY

Rotonda Tatiana
Università di Roma
Dip. Ingegneria Strutturale e Geotecnica
Via Monte d'Oro 28
00186 Roma
ITALY

Rotundi Leonardo
Gruppo Dipenta Costruzioni S.p.A.
Via Agrigento 5
00161 Roma
ITALY

Sainati Francesco
Edison S.p.A.
Via Rosellini 15/17
20124 Milano
ITALY

Santarelli Frederic
AGIP
P.O. Box 12069
20100 Milano
ITALY

Scavia Claudio
Politecnico di Torino
Dip. Ingegneria Strutturale
C.so Duca degli Abruzzi 24
10129 Torino
ITALY

Scesi Laura
Politecnico di Milano
D.S.T.M.
Piazza Leonardo da Vinci 32
20133 Milano
ITALY

Schiatti Luigi
A.E.C. Etschwerke
Via Dodiciville 8
39100 Bolzano
ITALY

Scroffi Claudio
Consorzio Lar
Via Melinotto, Loc. Vetta di Pegli
16100 Genova
ITALY

Segalini Andrea
Politecnico di Torino
Viale Martiri della Libertà 47
43100 Parma
ITALY

Sfondrini Giuseppe
Università di Milano, Dip. Scienze della Terra
Via Mangiagalli 34
20133 Milano
ITALY

Simoni Alessandro
Università di Bologna, Dip. Scienze della Terra
Via Zamboni 67
40126 Bologna
ITALY

Sterlacchini Simone
Università degli studi di Milano
Via Mangiagalli 34
20133 Milano
ITALY

Sterpi Donatella
Politecnico di Milano
Piazza Leonardo da Vinci 32
20133 Milano
ITALY

Stragapede Francesco
Via Vecchia P.le Montalbano 88/C
51034 Casalguidi PT
ITALY

Tobaldini Marco
Borini Costruzioni
Via Bellini 2
10121 Torino
ITALY

Tommasi Paolo
C.N.R. - C.S. Geologia Tecnica
c/o Fac. Ingegneria
Via Eudossiana 18
00184 Roma
ITALY

Toni Giancarlo
Università di Bologna
Via Zamboni 67
40127 Bologna
ITALY

Tonon Fulvio
Vicolo S. Antonino A3
31100 Treviso
ITALY

Totaro Liliana
Università di Roma
Dip. Ingegneria Strutturale e Geotecnica
Via Monte d'Oro 28
00186 Roma
ITALY

Valore Calogero
Università di Palermo
Dip. Ingegneria Strutturale e Geotecnica
Via Rosina Anselmi 11
90135 Palermo
ITALY

Varosio Giovanni
D'Appolonia
Via San Nazaro 19
16145 Genova
ITALY

Vicenzetto Tiziano
Vicenzetto S.r.l.
Via Municipio 18
35040 Villa Estense PD
ITALY

Xu Shulin
Geodata S.p.A.
C.so Duca degli Abruzzi 48/E
10129 Torino
ITALY

Zaninetti Attilio
Enel S.p.A. CRIS
Via Ornato 90/14
20162 Milano
ITALY

Zicarelli Maurizio
Università di Catania
Via Cardili 3
87030 Carolei CS
ITALY

Ziccarelli Maurizio
Università di Palermo
Via Silvio Pellico 10
87036 Rende CS
ITALY

Aydan Omer
Tokai University
Orido 3-20-1
424 Shimizu
JAPAN

Akimoto Masahiro
Kozo Keikaku Eng. Inc
4-38-13 Hon-cho
Nakano-Ku
164 Tokyo
JAPAN

Baba Masahiro
Kyoto University
Dept. of Mineral Sciences & Technology
Yoshida Honmachi Sakyo-ku
606 Kyoto
JAPAN

Denda Atsushi
Shimizu Corporation
Seavans South
No. 2-3 Shibaura 1-Chome
Minato-ku 10507 Tokyo
JAPAN

Ichikawa Yasuaki
Nagoya University
Dept. of Geotechnical & Environmental Engineering
Chikusa - 46401 Nagoya
JAPAN

Igari Tetsuo
Hazama Corporation
2-5-8 Kita Aoyama
Mianto-ku - 107 Tokyo
JAPAN

Ishikawa Koji
Chuo-Kaihatsu Corporation
3-13-5 Nishi-Waseda
Shinjuku-ku
169 Tokyo
JAPAN

Kikuchi Kohkichi
Kyoto University
Dept. Mineral Sciences & Technology
Yoshida Honmachi Sakyo-ku
606 Kyoto
JAPAN

Kitagawa Takashi
Nishimatsu Construction Co. Ltd
1-20-10 Toranomon
Minato-ku - 105 Tokyo
JAPAN

Kusumi Harushige
Kansai University
Dept. of Civil Engineering
564 Suita Osaka
JAPAN

Mito Yoshitada
Kyoto University
Dept. Mineral Science & Technology
Yoshida Honmachi
Sakyo-ku 606 Kyoto
JAPAN

Nishida Kazunori
Oyo Corporation
66-2 Miyaharacho 1 Chome
330 Omiya Saitama
JAPAN

Sakurai Shunsuke
Kobe University
Dept. of Civil Engineering
657 Kobe
JAPAN

Sassa Koichi
Kyoto University
Faculty of Engineering
Yoshida Honmachi Sakyo-ku
60601 Kyoto
JAPAN

Seto Masahiro
Nire
16-3 Onogawa
305 Tsukuba City
JAPAN

Shin Koichi
Central Research Institute
Electric Power Industry
1646 Abiko Abiko-shi
27011 Chiba ken
JAPAN

Tanaka Soichi
Oyo Corporation
2-6 Kudan-Kita 4 Chome
Chiyodaku Tokyo 102
JAPAN

Tani Kazuo
Central Research
Institute of Electric Power Industry
1646 Abiko Chiba
27011 Abiko City
JAPAN

Tanimoto Chikaosa
Kyoto University
School of Civil Engineering
Sakyo, Kyoto 606
JAPAN

Uehara Yoshihisa
Kyoto University
Dept. Mineral Science & Technology
Yoshida Honmachi
Sakyo-ku, Kyoto 606
JAPAN

Uemura Takeshi
Niigata University
Kamitokoro 1-12-B430
950 Niigata
JAPAN

Yokota Yasuyuki
CTI Engineering Co. Ltd
18 Fukuoka Bldg., 1-10 Watanabe Dori
2 Chome Chuo-ku, 810 Fukuoka
JAPAN

Yoshinaka Ryunoshin
Saitama University
255 Shimo Okubo
Urawa Shi
338 Saitama
JAPAN

Lee Chan Woo
Kolon Construction & Engineering Company
53-8 Chungdam-Dong
Kangnam-gu
135100 Seoul
KOREA

Lee Hi-Keun
Seoul National University
San 56-1 Sinrim Dong, Kwanak-ku
151742 Seoul
KOREA

Park Hyeong-Dong
Pai-Chai University
Dept. of International Resources Development
Doma 2 Dong Seo-gu
302735 Taejon
KOREA

Lamas Luis
L.E.C.M.
Rua da Sé 22
Macau
MACAU

Richards Laurie
Lincoln University
Dept. of Natural Resources Engineering
P.O. Box 84
Canterbury
NEW ZELAND

Barton Nick
NGI
P.O. Box 3930
Ullevaal Hageby
0806 Oslo
NORWAY

Borgerud Line Kristine
Statoil
G-4222
4035 Stavanger
NORWAY

Gutierrez Marte
NGI
P.O. Box 3930
Ulleval Hageby
0806 Oslo
NORWAY

Holt Rune M.
NTNU & IKU
IKU Petroleum Research
7034 Trondheim
NORWAY

Horsrud Per
IKU Petroleum Research
7034 Trondheim
NORWAY

Papamichos Euripides
IKU
Petroleum Research
7034 Trondheim
NORWAY

Risnes Rasmus
Stavanger College
P.O. Box 2557
Ullandhaug
4004 Stavanger
NORWAY

Singelstad Arne
Statoil
G4216
4035 Stavanger
NORWAY

Sonstebo Eyvind
IKU Petroleum Research
7020 Trondheim
NORWAY

Bogdanska Joanna
University of Warsaw
Faculty of Geology
93 Zwirki Iwigury Str.
02089 Warsaw
POLAND

Kwasniewski Marek
Silesian Technical University
Ul. Akademicka 2
44100 Gliwice
POLAND

Barros José
Junta Autonoma de Estradas
Praca da Portagem
2800 Almada
PORTUGAL

Castel Branco Falcao Joao Manuel
LNEC
Av: do Brasil 101
1799 Lisboa
PORTUGAL

Dinis da Gama Carlos
Instituto Superior Tecnico
Av. Rovisco Pais
1096 Lisboa
PORTUGAL

Grossmann Nuno Feodor
LNEC
Av. do Brasil 101
1799 Lisboa Codex
PORTUGAL

Lemos José
LNEC
Av. do Brasil 101
1799 Lisboa
PORTUGAL

Pinto da Cunha Antonio
LNEC
Av. do Brasil 101
1799 Lisboa Codex
PORTUGAL

Rodrigues José Delgado
ISRM c/o LNEC
Av. do Brasil 101
1799 Lisboa
PORTUGAL

Chafarenko Evgueni
STC Podzemgazprom
6-14 Parkovaya Str.
105203 Moscow
RUSSIA

Kovalenko Anatoly
KMA
Scientific Introduction Centre
Ul. Gorkogo 56-A
308031 Belgorod
RUSSIA

Novikov Anatoly
Russian Committee of Metallurgy
Slavyanskaya So 2/5
103718 Moscow
RUSSIA

Ocheretin Pavel
Ministry of Science of Russia
Moscow
RUSSIA

Rechitski Vladimir
Hydroproject Institute
Volokolamskoe Smosse 2
125812 Moscow
RUSSIA

Syrnikov Nikolai
Institute for Dynamics of Geospheres of RAS
Leninski Prospekt 38 - Korpus 6
117334 Moscow
RUSSIA

Likar Jakob
IRGO
Institute Mining Geotechnology & Environment
Slovenceva 93
1000 Ljubljana
SLOVENIA

Sovinc Ivan
Slovenian Geotechnical Society
Pod Kostanji 44
1000 Ljubljana
SLOVENIA

Suligoj Mitja
IRGO
Institute for Mining Geotechnology & Environment
Slovenceva 93
1000 Ljubljana
SLOVENIA

Erasmus Barend Jacobus
Dept. of Mineral & Energy
Private Bag X5
2017 Braamfontein
SOUTH AFRICA

Guler Gokhan
CSIR Mining Technology
P.O. Box 91230
2006 Auckland Park
SOUTH AFRICA

Jager Tony
C.S.I.R. Miningtek
P.O. Box 91230
Auckland Park
2006 Johannesburg
SOUTH AFRICA

Kotze Theunis
BKS Hatch
P.O. Box 62694
2107 Marshalltown
SOUTH AFRICA

Kuijpers Jan
CSIR Mining Technology
P.O. Box 91230
Auckland Park
2006 Johannesburg
SOUTH AFRICA

Leach Tony
Itasca Africa Pty Ltd
Box 38425 Booysens
2016 Johannesburg
SOUTH AFRICA

Legge Francis
Land Afrikaans University
Lab. for Materials
P.O. Box 524
2006 Auckland Park
SOUTH AFRICA

Pretorius Martin
Gengold
P.O. Box 55018
Eerstemyn 9466 R.S.A.
9459 Welkom
SOUTH AFRICA

Rymon Lipinski Wlodzimierz
Dept. of Mineral & Energy Affairs
Private Bag X59
0001 Pretoria
SOUTH AFRICA

Stewart Roger
CSIR Mining Technology
P.O. Box 91230
Auckland Park
2006 Johannesburg
SOUTH AFRICA

Van Der Merwe Jacob Nielen
Sasol Mining
Box 5486
2000 Johannesburg
SOUTH AFRICA

Webber Stephen
CSIR Mining Technology
P.O. Box 91230 - Auckland Park
2006 Johannesburg
SOUTH AFRICA

Romana Manuel
Polytechnical University of Valencia
Camino de Vera s/n
46071 Valencia
SPAIN

Alestam Mats
Sydkraft Konsult AB
20509 Malmo
SWEDEN

Ask Maria
Royal Institute of Technology
Engineering Geology
10044 Stockholm
SWEDEN

Brantmark Johan
Royal Institute of Technology
Dept. of Soil and Rock Mechanics
10044 Stockholm
SWEDEN

Cesano Daniele
KTH
Dept. Civil & Environmental Engineering
Div. of Land & Water Resources
10044 Stockholm
SWEDEN

Glamheden Rune
Chalmers University of Technology
Dept. of Geotechnical Engineering
41296 Goteborg
SWEDEN

Hakami Eva
Itasca Geomekanik AB
Stenbocksgatan 1
11430 Stockholm
SWEDEN

Jing Lanru
Royal Institute of Technology
Division of Eng. Geology KTH
10044 Stockholm
SWEDEN

Klasson Hans
Vattenfall Hydropower AB
P.O. Box 50120
97324 Lulea
SWEDEN

Lanaro Flavio
Royal Institute of Technology
Civil & Environmental Engineering - KTH
10044 Stockholm
SWEDEN

Larsson Harry
Rock Store Design - ROX AB
Dorjgatan 7
13343 Saltsjobaden
SWEDEN

Olsson Lars
Gestatistik AB
Box 116
14722 Tumba
SWEDEN

Olsson Olle
SKB
P.L. 300
57295 Figeholm
SWEDEN

Pusch Roland
Geodevelopment AB
Ideon Research Park
22370 Lund
SWEDEN

Ranqvist Gunnar
SKB Aspo Lab.
Pl 300
57295 Figeholm
SWEDEN

Ringstad Cathrine
IKU Petroleum Research
7034 Trondheim
SWEDEN

Stenberg Leif
SKB
Aspo Hard Rock Laboratory
PL 300
57093 Figeholm
SWEDEN

Stephansson Ove
Royal Institute of Technology
10044 Stockholm
SWEDEN

Sturk Robert
Royal Institute of Technology
Soil & Rock Mechanics
10044 Stockholm
SWEDEN

Bonzanigo Luca
Viale Stazione 16/A
P.O. Box 1152
6501 Bellinzona
SWITZERLAND

Brox Dean
Amberg Consulting Engineers Ltd
Trockenloostrasse 21
8105 Regensdorf-Watt
SWITZERLAND

Egger Peter
Rock Mechanic Lab. EPFL
Politecnico di Losanna
1015 Lausanne
SWITZERLAND

Epars Pierre
Bonnard & Gardel Ingenieurs - conseils SA
Av. de Cour 61
P.O. Box 241
1001 Lausanne
SWITZERLAND

Guglielmini Gabriele
Geobrugg
P.O. Box 239
6595 Riazzino-Locarno
SWITZERLAND

Palmoso Stefano
Paler S.A.
Piazza Brocchi 5
6926 Montagnola
SWITZERLAND

Peter Corrado
Prader AG, Postfach 6839
Waisenhausstrasse 2
8023 Zurich
SWITZERLAND

Steiner Walter
Balzari & Schudel AG
Muristrasse 60
3000 Bern 16
SWITZERLAND

Watson Timothy
C.E.R.N.
St. Division
1211 Geneva 23
SWITZERLAND

Chang Wen Cherng
National Expressway Eng. Bureau
4Fl. No.21Lane 127 Sec. 2
Ming-Sheng E. Road
104 Taipei
TAIWAN

Ou Chin-Der
Public Construction Commission
Executive Yuan
9/Fl. No.4 Chung-Hsiao
W.Rd.Sec.1
100 Taipei
TAIWAN R.O.C.

Harnpattanapanich Thanu
Royal Irrigation Dept.
40/22 Soi Srisategn
Patchakasam Rd.
Sampran City
THAILAND

Trinetra Kijja
R.I.D. Geotechnical Division
40/22 Soi Srisategn
Patchakasam Rd.
Sampran City
THAILAND

Bilgin Nuh
Istambul Technical University
Maden Fakultesi
Maslak
80626 Istambul
TURKEY

Genis Melih
Karaelmas University
Dept. of Mining Engineering
67100 Zonguldak
TURKEY

Gercek Hasan
Karaelmas University
Dept. of Mining Engineering
67100 Zonguldak
TURKEY

Ozgenoglu Abdurrahim
Turk Ulusal Kaya, Mekanigi Dernegi
Maden Muh Bolumu O.D.T.U.
06531 Ankara
TURKEY

Senyur M. Gurel
Hacettepe University
Engineering Faculty, Dept. of Mining
Beytepe Ankara
TURKEY

Arnold Peter
Camborne School of Mines
Redruth Cornwall
TR153SE Redruth
UNITED KINGDOM

Davies Nicholas
UK Nirex Ltd.
Curie Avenue Harwell Didcot
OK11 ORH Oxfordshire
UNITED KINGDOM

Emsley Simon
Golder Associates Ltd
54 Moorbridge Road
SL68BN Maidenhead
UNITED KINGDOM

Fang Z.
Imperial College, Dept. of Earth Resources
Engineering, Prince Consort Road
SW72BP London
UNITED KINGDOM

Gordon Terry
Ove Arup & Partners
Cambrian Buildings, Mountstuart Square
CF16QP Cardiff
UNITED KINGDOM

Gunn David
British Geological Survey
Keyworth
NG125GG Nottingham
UNITED KINGDOM

Harrison J.P.
Imperial College
Dept. of Earth Resources Engineering
Prince Consort Road
SW7 2BP London
UNITED KINGDOM

Hope Victoria
University of Surrey
Dept. of Civil Engineering
Surrey - GU25XH Guildford
UNITED KINGDOM

Hudson John Anthony
7 The Quadrangle
AL86SG Welwyn Garden City
UNITED KINGDOM

Jiao Yong
Rock Engineering Consultants
16 Devonshire Road
Colliers Wood
SW192EN London
UNITED KINGDOM

Kapenis A.
Imperial College
Dept. of Resources Engineering
Prince Consort Road
SW72BP London
UNITED KINGDOM

Lu Ping
Queen Mary & Westfield College
Geomaterials Unit
Dept. of Engineering
E14NS London
UNITED KINGDOM

Millar Dean L.
Imperial College
Dept. of Earth Resources Engineering
Prince Consort Road
SW72BP London
UNITED KINGDOM

Murdie Ruth
Applied Seismology & Rock Physics Laboratory
Dept. of Earth Sciences
Keele University
STS 5BG Keele Staffs
UNITED KINGDOM

Yeo I.
Imperial College
Dept. of Earth Resources Engineering
Prince Consort Road
SW72BP London
UNITED KINGDOM

Pine Robert
Golder Associates Ltd
54-70 Moorbridge Road
SL68BN Maidenhead
UNITED KINGDOM

Reid T.R.
Imperial College
Dept. of Earth Resources Eengineering
Prince Consort Road
SW72BP London
UNITED KINGDOM

Sharp John
Geo Engineering
Coin Varin St. Peter
JE3 7EH Jersey
UNITED KINGDOM

Sloan Andrew
W.A. Fairhurst & Partners
11 Woodside Terrace
G37XQ Glasgow
UNITED KINGDOM

Stimpson Ian
Keele University
Applied Seismology & Rock Physics Laboratory
Dept. of Earth Sciences
STS5BG Keele Staffs
UNITED KINGDOM

Wei Lingli
Golder Associates Ltd
Landmere Lane, Edwalton
NG124DG Nottingham
UNITED KINGDOM

Young Paul
Keele University
Applied Seismology Rock Phisics Laboratory
Dept. of Earth Sciences
STS 5BG Keele Staffs
UNITED KINGDOM

Zimmerman Robert
Imperial College
Ere Dept.
SW72BP London
UNITED KINGDOM

Abousleiman Younane
The Rock Mechanic Institute
Sec.T109 - 100 East Boyd
73019 Norman
U.S.A.

Bartholomew Jennifer
ISRM
2112 Drew Avenue South
55416 Minneapolis MN
U.S.A.

Board Mark
Itasca Consulting Group
708 South Third Street
55415 Minneapolis MN
U.S.A.

Dershowitz William
Golder Associates
4104 148th Avenue
98052 Redmond WA
U.S.A.

Einstein Herbert
MIT Room 1-342
02139 Cambridge MA
U.S.A.

Fairhurst Charles
University of Minnesota
500 Pillsbury Drive SE
55455 Minneapolis
U.S.A.

Fairhurst Charles E.
MTS Systems Corporation
14000 Technology Drive
55344 Eden Prairie MN
U.S.A.

Garagash Dmitri
University of Minnesota
Civil Engineering Dept.
500 Pillsbury Drive S.E.
55455 Minneapolis MN
U.S.A.

Germanovich Leonid
Oklahoma University
100 E. Boyd - P119
73019-1014 Norman OK
U.S.A.

Goodman Richard E.
University of California
Davis Hall
94720 Berkeley CA
U.S.A.

Haimson Bezalel
University of Wisconsin
1509 University Avenue
53706 Madison WI
U.S.A.

Kim Kunsoo
Columbia University
812 S.W. Mudd
10027 New York NY
U.S.A.

Markham John
Itasca Consulting Group Inc.
708 South Third Street
Suite 310
55415 Minneapolis MN
U.S.A.

Rhett Douglas
Phillips Petroleum Company
132 GB
Phillips Reasearch Center
74004 Bratlesville OK
U.S.A.

Roegiers Jean-Claude
University of Oklahoma
Rock Mechanics Institute
Energy Center
730190628 Norman
U.S.A.

Santos Helio
Petrobras
418 Rosewood Drive
73069 Norman OK
U.S.A.

Smeallie Peter
American Rock Mechanics Assn.
600 Woodland Terrale
22302 Alexandria VA
U.S.A.

Students from the Politecnico di Torino
(Rock Mechanics and Rock Engineering Group)

AIASSA Santina
AVAGNINA Nicola
BALLERI Roberta
BARLA Marco
BRIGNONE Giuliano
CARPEGNA Massimo
CORBO Giusi
FELICE Alessandro
GARIZIO Germana
PESCARA Moreno
PIASSO Guido
SCHIAVINATO Luca

Eurock '96, Barla (ed.)© 2000 Balkema, Rotterdam. ISBN 90 5410 843 6

Author index
Index des auteurs
Autorenverzeichnis